Humanizing Artificial Intelligence

Humanizing Artificial Intelligence

Psychoanalysis and the Problem of Control

Edited by
Luca M. Possati

DE GRUYTER

ISBN 978-3-11-100736-6
e-ISBN (PDF) 978-3-11-100756-4
e-ISBN (EPUB) 978-3-11-100759-5

Library of Congress Control Number: 2023941096

Bibliographic information published by the Deutsche Nationalbibliothek
The Deutsche Nationalbibliothek lists this publication in the Deutsche Nationalbibliografie; detailed bibliographic data are available on the Internet at http://dnb.dnb.de.

Cover image: kentoh/iStock/Getty Images Plus
Printing and Binding: CPI books GmbH, Leck

www.degruyter.com

Table of Contents

Luca M. Possati

Introduction

This book intends to collect a series of contributions for an interpretation of artificial intelligence from a psychoanalytic point of view. It claims neither to define a method nor to draw universal conclusions about the nature of technology or the human mind. At its core there is the analysis of the human-technology relationship and its further relationship with the phenomenon of technological innovation. It is therefore a highly interdisciplinary work and intends to address researchers, students, and the public interested in these issues.

Why do we need a psychoanalytically based approach to AI? How can a discipline, or rather a set of doctrines that are very different from each other and lacking in methodological unity, tell us something significant about the nature of the human being and his relationship with technology? This book does not intend to defend the point of view of psychoanalysis, or of a specific psychoanalytic school. Instead, this book intends to open a series of *explorations* on the statute of psychoanalysis, and more generally on the concept of the unconscious and on its transformations in relation to new digital technologies and AI.

The need for this research lies in a new interpretation of the classic problem of the control of technology. It is evident that the earlier assumption that technological evolution would automatically lead to significant social and human progress can no longer be sustained today. The ambivalence of technology has become a standing topic in public, philosophical, and scientific debates. The scientific discussion about how to acquire and establish orientational knowledge for decision-makers facing the ambivalence of technology is divided into two branches: the ethics of technology and technology assessment (Grunwald 1999, 2018). These two branches are based on different assumptions concerning how to orient technology policy: the philosophical ethics branch, of course, emphasises the normative implications of decisions related to technology and the importance of moral conflicts, while the technology assessment branch relies mainly on sociological or economic research.

The problem of evaluating and controlling technological development is at the heart of the so-called "Collingridge dilemma," which can be formulated as follows:

> "attempting to control a technology is difficult, and not rarely impossible, because during its early stages, when it can be controlled, not enough can be known about its harmful social consequences to warrant controlling its development; but by the time these consequences are apparent, control has become costly and slow." (Collingridge 1980, 19)

https://doi.org/10.1515/9783111007564-001

For David Collingridge, technological development is always faster than the ability to understand its social effects. This asymmetry creates a strange effect: when changing a technology is simpler, especially at the beginning of its development, it is not perceived as a necessity, but when change is perceived as a necessity, it is no longer simple—it has become expensive and dangerous. "It is just impossible to foresee complex interactions between a technology and society over the time span required with sufficient certainty to justify controlling the technology now, when control may be very costly and disruptive" (Collingridge 1980, 12).

It's important to note that the dilemma is not about technological development itself but about the perception that humans have of it and the awareness of its limits and effects. In fact, scholars underline that the technological development we produce exceeds our level of awareness and knowledge, and this affects our ability to forecast the social implications of technology: "A technology can be known to have unwanted social effects only when these effects are actually felt" (Collingridge 1980, 14). Why is it that, as technologies develop and become diffused, they become ever more resistant to controls that seek to alleviate their unwanted social consequences? To solve the dilemma, Collingridge (1980) develops a reversible, flexible decision-making theory that can be used when the decision-maker is still ignorant of the effects of a technology. According to Collingridge, the essence of the control of technology is not in forecasting its social consequences "but in retaining the ability to change a technology, even when it is fully developed and diffused, so that any unwanted social consequences it may prove to have can be eliminated or ameliorated" (20–21). The important thing is to understand how to make the decisions that influence technological development in such a way as not to remain prisoners of them.

Now, there are different interpretations of the problem of control. Some are markedly alarmist and refer to the concept of singularity (Kurzweil 2005). In this regard, the example of Bostrom (2014) and Tegmark (2017) can be cited. The problem of control is then interpreted as the problem of how to prevent superintelligence from harming human beings. However, this interpretation risks fueling apocalyptic visions and excessive concerns.

There is also another interpretation of the control problem which abandons an alarmist tone and focuses more on the human-machines relationship and its potential. As Russell (2019) claims, "If we build machines to optimize objectives, the objectives we put into the machines have to match what we want, *but we do not know how to define human objectives completely and correctly*" (170, emphasis added). Human beings put their goals into the machine, but this is exactly the problem. Humans want the machine to do what we want, "*but we do not know how to define human objectives completely and correctly,*" and we often act in ways that contradict our own preferences. Humanity is not a single, rational entity but "is composed of nasty, envy-driven, irrational, inconsistent, unstable, computationally limited, complex, evolving, heterogeneous entities. Loads and loads of them" (211).

The main challenge is to understand the nature of our goals and preferences. In Russell's (2019) view, "Preference change presents a challenge for theories of rationality at both the individual and societal level. . . . Machines cannot help but modify human preferences, because machines modify human experience" (241). How can we communicate our needs, values, and preferences to AI systems? This is a crucial problem in our world, where the influence of AI-based technologies is growing enormously. Unconscious dynamics influence AI and digital technology in general, and understanding them is essential to ensuring that we have better control of AI systems. For this reason, studying the way in which technology influences and orients our emotional and cognitive unconscious is a crucial undertaking to ensure a balanced relationship between human beings and technology.

In the first chapter, Paul Jorion develops an analysis of the concepts of artificial consciousness (AC) and artificial general intelligence (AGI). He claims that the connection between them is misguided as it is based on a folk psychology representation of consciousness. There exists, however, a path leading from AI to AGI which skips entirely the need to develop AC as a stepping-stone in that direction; that path inspired by Freud's metapsychology sets at the core of the human mind a network of memory traces acted by an affect dynamics.

In Chapter 2, starting with Freud's concept of the psychic machine, Hub Zwart discusses Lacan's effort to elaborate on this view with the help of 20th-century research areas (computer science, linguistics, cybernetics, molecular biology, etc.), resulting in the famous theorem that the unconscious is structured as a language. Subsequently, two closely related questions are addressed, resulting from a mutual encounter between psychoanalysis and AI, namely: How can psychoanalysis contribute to coming to terms with AI and to what extent does AI allow us to update psychoanalytic theories of the unconscious?

In Chapter 3, Kerrin A. Jacobs claims that AI companionship promises a new way of coping with loneliness experiences in highly digitalised societies. In a first step some basic criteria that characterise the relationship with a companion AI (social x-bots) as distinct from human relatedness are sketched. While AI companionship is often praised for the potential to cope with loneliness its crucial flaw is its lacking of an intersubjective dimension, which is essential for the human condition. The central hypothesis is that AI companionship cannot solve the problem of loneliness, which is elaborated on in a second step.

In Chapter 4, Andre Nusselder analyses decision-making by football referees supported by Video Assistance Referee (VAR) technologies, with the goal of implementing AI so that accurate and fair decisions can be made without interrupting the flow of the game too much. This chapter analyses the connection between these technologies and affective self-regulation of those involved. It does so from Norbert Elias's theory of civilisation in which he analyses – using Freud's metapsychology – how increased civilised behaviour leads to increased self-control of individuals. The chapter argues that the aim of making football a fairer game with the use of AI has a similar impact on those involved and is thus a next step in the movement towards the posthuman condition which takes place subtly as humans adapt to it without generally being conscious of it but with far-reaching effects.

In the last chapter, Roberto Redaelli clarifies the moral status of AI systems by applying to them the notion of moral mediator developed by P. P. Verbeek in the field of Science and Technology Studies (STS). More precisely, we propose to define artificial intelligent systems as moral mediators of a particular kind, i. e. as possessing technological intentionality linked to composite intentionality. To this end, it is first necessary to show that the common view of technology held by various forms of instrumental theories is insufficient for the purpose of understanding the agency of AI systems. Redaelli analyses some paradigmatic positions that assign a certain moral status to technological artefacts, such as those of D. G. Johnson and J. Sullins, in order to compare them with Verbeek's postphenomenological approach.

Paul Jorion

A Freudian Implementable Model of the Human Subject

Abstract: This chapter aims to define a new model of AI from Freudian metapsychology. The main thesis is that, contrary to common assumption within the artificial intelligence community, help will not come from techniques still to be developed aiming at building an "artificial general intelligence," aka "machine common sense," but from a better model of what is a human subject. What needs to be implemented in the robot is a simulation of the mechanism allowing a human subject to acquire instead of an "artificial general intelligence" a "common moral sense" such as that builds over the years in the child and then in the adolescent. The computer solutions to do so are already available.

Keywords: metapsychology, human subject, artificial general intelligence

Introduction

For 80 years now speculative thinkers have debated Isaac Asimov's "Three Laws of Robotics", a bundle of three simple interlocking directions supposedly sufficient for regulating the behaviour of robots and making their daily interaction with human beings both useful and unproblematic.

Although Asimov's Three Laws have been the centre of profuse and vivid exchanges, they'd been entirely ignored when engineers started to implement actual robots, or "intelligent machines," broadly speaking.

The reason for such a surprising disconnect is actually straightforward: Asimov's robots are autonomous while the actual robots engineered up to now are at best semi-autonomous only: they're only given a free hand whenever their capabilities clearly exceed ours, with the ultimate decision-making remaining ours.

But this semi-autonomous status will only last as long as our decision-making remains more efficient than the robots' own. As soon as that ceases to be the case, full autonomy will no doubt be granted them.

Contrary to common assumption within the artificial intelligence community, help will not come from techniques still to be developed aiming at building an "artificial general intelligence," aka "machine common sense," but from a better model of what is a human subject.

What needs to be implemented in the robot is a simulation of the mechanism allowing a human subject to acquire instead of an "artificial general intelligence" a

https://doi.org/10.1515/9783111007564-002

"common moral sense" such as that builds over the years in the child and then in the adolescent. The computer solutions to do so are already available.

An autonomous robot is out of necessity of a Freudian concept; otherwise, it will never be more than Microsoft's ill-fated TAY: a moron that is easily convinced to become sexist and racist after a dozen hours of conversation only with users.

Microsoft's TAY: the damages of an AI deprived of a personal history

In 2016, Microsoft released a piece of software able to carry on conversations with users: a *chatbot.* That experiment actually duplicated a project previously released to great success in China by the same IT giant, called Xiaoice, a venture that was deemed most impressive as it had held over 40 million conversations with users. TAY stood for "Thinking About You".

At the end of 16 hours only, Microsoft was forced to stop the experiment as TAY was not behaving: it relished in sexist and racist jokes. When asked about the Holocaust, it claimed it was bogus and that it had never taken place, along with, displaying to emphasise the point, a jolly hand-clapping emoji. That had to be stopped. Prompted again some time later, TAY boasted that it had smoked weed, that that had made it very happy, and that it had been done in full display of cops ("I'm smoking kush infront the police").

What had happened? Facetious users had encouraged TAY to state such outrage. What did it reveal? It revealed that TAY had no personality of its own and that it was but exchanges with its users that allowed it to build the likeness of a personal history. It goes without saying that this is not the way it should work out, and some earlier artificial intelligence projects had of course thought out that kind of issue.

Isaac Asimov's "Three Laws of Robotics"

There exists a bundle of principles labelled the "Three Laws of Robotics," having been initially formulated in the early 1940s by Isaac Asimov (1920–1992), a highly regarded science-fiction writer. Had TAY followed the Three Laws of Robotics, what happened in life would never have taken place. A paradox lies here, which is this: if you think of programmers in artificial intelligence and in computer science generally speaking, those are people who are among the most dedicated readers of science fiction, the most interested in that literary genre as they belong to that part of the population called "nerds", and more specifically "geeks", computer specialists who, in fact, have few activities apart from interacting with their computers or playing video games. And it's very curious that people who are so knowledgeable about the discussions that have taken place around those "Three Laws of Robotics" have passed up a piece of software that in fact completely ignores the lessons learned during that very important debate dating back to the early 1940s. What's this? Eighty-two years of debate around the "Laws of Robotics" and still the pathetic sinking of the TAY adventure?

Isaac Asimov was born in Russia in 1920. He died in the United States in 1992. In academia, he was a professor of biochemistry at Boston University, but he is known as one of the greatest science-fiction writers ever. Asimov started writing in the very early 1940s, in particular around this theme of the Laws of Robotics, i. e., the principles that robots should respect in their interactions with human beings. He developed the theme little by little in his works, thinking about it as he went along. At first, in his very first short stories, these laws of robotics were implicit. Then he started to express them explicitly. Other science-fiction writers invoked them in their writings and a kind of general discussion took place. After Asimov's death in 1992, the process went along: some writers came back to this and introduced new laws of robotics, staged new paradoxical developments of them, etc.

Here are Asimov's "Three Laws of Robotics":

1. A robot may not injure a human being or, through inaction, allow a human being to come to harm.
2. A robot must obey the orders given to it by human beings except where such orders would conflict with the First Law.
3. A robot must protect its own existence as long as such protection does not conflict with the First or Second Law.
   Handbook of Robotics,
   56th Edition, 2028 A.D.
   (Asimov 1950: 8).

Later on, a Fourth Law was added, and from then on the so-called "Zero Law", which is that a robot is not to act in any manner that would endanger humankind as a whole.

In discussions that took place in some of Asimov's later texts, he made it clear in relation to that Zero Law that it is extremely difficult to respect since it requires a global view, a reflection obliging one to have an overall representation of humankind independently of who its different representatives are. To possibly endanger the existence of particular human beings in the name of the human race as a whole requires in fact a "meta-"principle, that is to say, that it is unlike the

Laws of Robotics, which Asimov imagines to be simply what we call algorithms: the way he phrases it is "mathematical procedures". With the Zero Law of protecting the whole of humankind, here is indeed something of a higher level since no simple algorithm can implement such a thing.

The literature that would develop over the years shows the full set of contradictions arising from these simple laws. For example, contradictions emerge by merely interchanging the order wherein the different laws are called up in a reasoning. There are plenty of occurrences of ambiguity: in order to apply those three laws, the robot must somehow hold a prescient vision of the future, e. g. if it is told to operate on someone because the person will die if the operation doesn't take place and it sticks to its pure and simple principle of not hurting a human being, it will abstain from performing it, and so on.

Asimov's robots are autonomous

So there you have it: 82 years of discussions. Eighty-two years until the invention of TAY: discussions about what is possible, what is impossible, "Can you imagine this or not?" etc.

What is fundamental of course in those laws of robotics is that we imagine robots making decisions on their own: that are autonomous. That is to say that they do not consult human beings before any of their moves, nor are they machines which are manipulated remotely when a human being is actually making decisions on their behalf. In the current environment, it needs to be recalled, the Three Laws of Robotics are not being applied as there are no autonomous robots as such.

For there to be an autonomous robot that respects ethical principles, it would have to be accountable, i. e. the opportunity would need to exist that it'd be brought before some judicial instance and punished for actions going against the standing legal framework. Such is not currently the case, however: the only robots existing today are of a type where we tell them what to do, leaving them initiative within a very narrow range only.

One thing also that had been noticed right away by ethicists was that research into robotics has been carried out from the very beginning in the military field, surroundings so defined that it was absolutely impossible to apply the "Three Laws of Robotics", starting with the First Law that no human being should be harmed, since the very principle of robotics in the military field is instead precisely that some human beings should be hurt, especially those threatening the further existence of the robot itself.

The very principle that a robot respects humans before it even thinks of protecting itself is also unrealistic and unenforceable since a robot is an expensive piece of machinery and it will be instructed to defend itself so that it is not easily destroyed, even if this means that in truly contentious cases it neutralises human beings threatening it, and other things of that order. Here is something by the way that didn't escape Asimov himself and he came up accordingly in one of his robot stories ("Risk", in 1955) with an alternative tongue-in-cheek "Three Laws":

"First Law: Thou shalt protect the robot with all thy might and all thy heart and all thy soul.

Second Law: Thou shalt hold the interests of U.S. Robots and Mechanical Men, Inc. holy provided it interfereth not with the First Law.

Third Law: Thou shalt give passing consideration to a human being provided it interfereth not with the First and Second Laws. (Asimov [1964] 1968: 139).

Robin Murphy's Laws of Robotics: guiding laws for human robot users

But as Microsoft's TAY project emphasised, working on autonomous robots happens to be very much in line with the notion of developing artificial intelligence altogether. It would only be contradictory if we excluded that robots would ever be autonomous, if they would continue to do only things that are specifically asked of them without displaying any sense of initiative. Robin R. Murphy and David D. Woods addressed that issue in a detailed manner in a 2009 article entitled: "Beyond Asimov: The Three Laws of Responsible Robotics". There, Murphy and Woods proposed to replace Asimov's Laws with something of an entirely different nature: laws about robots but applying to human beings designing robots, not applicable to the robots themselves.

The "Three Laws of Responsible Robotics" as phrased by Murphy and Woods are the following:

1. A human should not release a robot where the highest level of legal and professional implications has not been attained. "The highest professional ethics should also be applied in product development and testing" (Murphy and Woods 2009: 17).
2. A robot must meet the expectations of human beings according to the functions determined for it. "Robots must be built so that the interaction fits the relationships and roles of each member in a given environment" (Murphy and Woods 2009: 18)

3. A robot must have sufficient autonomy to protect its own existence to the extent that such protection allows for a smooth transfer between it and control by other agents, consistent with the First and Second laws. "Designers [should] explicitly address what is the appropriate situated autonomy (for example, identifying when the robot is better informed or more capable than the human owing to latency, sensing, and so on) and to provide mechanisms that permit smooth transfer of control" (Murphy and Woods 2009: 19)

What is Murphy and Woods's aim with those alternative three laws? They mean that in the context of our employment of robots, a robot must enjoy a certain amount of autonomy so that we human beings are able to take advantage of its superiority over us in certain domains, for instance, faster response time, greater power, and being able to perform operations that are physically difficult for human beings and that a robot can more easily realise. But should any danger arise, the robot must be able to transfer decision-making instantly to the human operator and vice versa. In other words, we must be in a situation where the robot should be autonomous as far as the qualities which are proper to it are concerned, namely those exceeding in a particular realm those of the human, but for the rest, it must be able to transfer back responsibility to a human being in a split second.

Those Three Murphy's Laws (not to be confused of course with the other more famous Murphy's Law: "a supposed law of nature, expressed in various humorous folk sayings, that anything that can go wrong will go wrong", according to Wikipedia) are compatible with the way we are operating at the moment. It is a way of reformulating Asimov's Three Laws of Robotics but in a context where the robot continues to be an aid to the human being and should only prolong the human to the extent that its capabilities exceed his.

The debate and ingenuity of Asimov himself and other debaters around his "Three Laws" have, however, revealed that his laws of robotics won't do the job as he himself graciously underlined in the short stories composing his two collections of robot stories entitled *I, Robot* (1950) and *The Rest of the Robots* (1964).

"Checking out" rules" such as the "Three Laws" can only go so far in making robots ethical, in the same way as laws are incapable on their own of making a human society viable. Indeed, humans need to stick to a large extent spontaneously to a virtuous behaviour before laws can provide a containing framework for trespassing excesses.

What needs therefore to be implemented now is a process that would induce in a robot something of the essence of virtue, meaning that a framework such as the "Three Laws" would only be there as a complement providing a final touch of control.

James H. Moor: Four Ways of Being Ethical for Robots

In that perspective of clarifying what would be a robot virtuous by concept, James H. Moor distinguishes in "Four Kinds of Ethical Robots" (2009) various degrees of moral assessment that a robot can offer. Moor is a professor of philosophy at Dartmouth College, an institution legendary of course in the artificial intelligence world for being the location where the overall artificial intelligence project was first outlined at a conference in 1956.

At the first degree of moral assessment, Moor calls "ethical agents", machines that are ethical in an entirely passive way for having as part of their design a feature protecting their users in some way or other. For example, a watch can be considered ethical insofar as it is equipped with an alarm alerting people that they need to perform a particular task.

The second degree of ethical assessment is to be found with "implicit ethical agents": those where security mechanisms have been purposely designed to protect users.

Those first two degrees are in line with a remark made in one of Asimov's own examinations of his "Three Laws", when he reminded that those laws are nothing more than general principles governing the operation of any machine or even tool. A machine or tool, he notes, must serve some specific purpose so as to be useful to human beings. Additionally, says Asimov, a machine or tool should present no danger to its user and there should ideally exist a mechanism that makes it stop at the moment when it can present a danger to human beings. Finally, it must be robust enough so that it does not break at the slightest use. In other words, those "Three Laws of Robotics" are merely derived from general principles applying to the functioning of any machine or tool: "Consider a robot, then, as simply another artefact. It is not a sacrilegious invasion in the domain of the Almighty, any more (or any less) than any other artefact is" (Asimov [1964] 1968: 14).

The third degree of moral assessment that a robot can offer is that of being an "explicit ethical agent". Such robots can recognise when particular laws or ethical principles are being infringed. This would involve one imagining an associated expert-system filtering certain types of behaviour according to major principles written into it.

The fourth degree of moral assessment is that of a robot which is strictly speaking "ethical". Moor calls those "full ethical agents". These are effectively autonomous robots that make all their decisions according to principles which are fully theirs, with no need to consult a table of instructions or directions provided by a supervising human being; in other words, those robots behave in an ethical way by nature, without having to exchange on a constant basis with a supervisor. Moor writes: "*full* ethical agents have those central metaphysical features that we usually attribute to ethical agents like *us* – features such as consciousness, intentionality and free will." Moor fails, however, to provide a recipe for how those "metaphysical features" might be acquired and it is here that a model of the human subject borrowed from Freud's metapsychology will turn out to be most useful.

TAY revisited: it had not gone mad, it had just joined the far right

The difficulties arising from Moor's "full ethical agents" are those that were truly embodied in TAY, Microsoft's chatbot, from which we immediately grasped that its understanding of the world was for a crucial part extracted from the conversations it held with users. In such a way that when it had to deal with facetious or far-right interlocutors it got easily persuaded that the solution to all evils was to get rid of the Jews, of the Arabs, and so on, and that the Holocaust, on the one hand, didn't happen and on the other hand, if it did, would have been a good thing, etc. Why did TAY say this? Of course, because there was absolutely nothing inside it as a matter of safeguards, of railings, in the way of filtering what it might say, having as a sole source of moral judgment whatever it had been told by users.

Why had the TAY approach actually worked with a product similar to TAY in the Chinese context? Probably because the Chinese setting is one of greater deference, greater respect for each other's business, and – it has to be said – also much quicker punishment for the bad guys. That didn't happen with a similar piece of software when it was released in the United States, in such a way that within a few hours, the software had to be taken down. It got restarted a little later and became then very sententious, saying things like: "There is no difference between men and women", etc. Those were clearly canned responses, i. e. words that were not the outcome of any "reasoning" by the AI but had been put there and were then retrieved at the right moment but in an absolutely mechanical way.

What would have been needed? First of all, a filter of an expert-system nature containing a set of rules and the ability for TAY to check sketches or drafts of what it was planning to say with some principles inscribed in it and not to let through what would contravene the corresponding set of rules.

Thinking about what happened, to sum up quickly, TAY had become a Trumpist. Indeed, in a response to Twitter user @icbydt TAY said, "bush did 9/11 and Hitler would have done a better job than the monkey we have now. donald trump is the only hope we've got."

Does that mean that Trumpists do not exist in the real world? Of course they do. To call things by their name, what we have here is a conflict between the people who make artificial intelligence, the conceivers and programmers, and the Trumpists and that's why there was an immediate outcry. Was TAY decried because no one in the world denies the existence of the Holocaust? Of course not, it's just that robot designers would like their products not to utter the kind of horrors proper to "those people" at the opposite end of the political spectrum.

It was James H. Moor, whom I just mentioned, who pointed out that in the case of Hurricane Katrina, a robot's response could hardly have been worse than that of the US Government's: "For instance, a robotic decision-maker might be more competent and less biased in distributing assistance after a national disaster like Hurricane Katrina, which destroyed much of New Orleans. In that case, the human relief effort was dangerously incompetent, and the coordination of information and distribution of goods was not handled well. In the future, ethical robots might do a better job in such a situation" (Moor 2009). If Moor doesn't mention TAY it is of course because his article predates by seven years Microsoft's release of its misfortunate chatbot.

So, it's not that people behaving like TAY don't exist; it's just that people designing robots would like to think that if a robot becomes a "thinking robot", it doesn't behave like the worst kind of scoundrel, even though not only the worst kind of scoundrel exist in the real world but also the same obnoxious reasoning may underly the reactions of an actual government, and in this case, government both at the local and federal levels.

A legal personality for robots? Not as long as they're not emotional

What is a current robot missing to be autonomous? First of all, it must be authorised to be so. Does that mean that it must be assigned a legal personality? The arguments for giving robots a legal personality have so far not been very convincing, in particular in light of the havoc that we see, wreaked as a consequence of attributing a legal personality to corporations, often leading to situations where the power of companies acting as mammoth individuals exceeds that of proper persons and human beings are crushed precisely by the power of corporations. So there is no compelling argument in favour of a legal personality for robots; the current context seems satisfactory enough where the responsibility for a robot's wrongdoing gets assigned, according to circumstances, to the maker or the user.

Nonetheless, a reasonable case can be made for the principle of an autonomous robot. And if the notion has been acknowledged, it won't be enough for there to be an expert-system simply sorting out how and in what precise order words should be uttered, just as a person's Super-Ego does in a psychoanalytical perspective, such as an occurrence in which, at the last instant before saying something, we tell ourselves, "Oops! As this man has a big nose, I'd better mention his mouth than his nose", and things of that nature. But above all, it would be essential for a speaking robot that the words that come spontaneously to its mind don't require immediate salvage and be replaced in an emergency mood by less offensive, more appropriate words.

How come that although this all springs to mind, it got ignored in TAY's case? Because the chatbot had been equipped with a broad lexicon of words that it could use, but there was no moral evaluation of how they would be retrieved. To call it by its name, there was no *affect dynamics* linked to any of the information stored within TAY. In such a way that the system was easily persuaded that what was required from it was to please at all costs the user, i. e. to ape the user's opinions and that, when he or she had had fun expressing Trumpist views that went against everything that is "politically correct", TAY would, however, make them its own in no time.

What should be concluded from this is that contrary to common assumption within the artificial intelligence community, help will not come from techniques still to be developed aiming at building an "artificial general intelligence," aka "machine common sense," but from a better model of what is a human subject.

It will be shown that what needs to be implemented in the robot is a simulation of the mechanism allowing a human subject to acquire instead of an "artificial general intelligence" a "common moral sense" such as that builds over the years in the child and then the adolescent. The computer solutions to do so are available by now.

An autonomous robot is out of necessity of a Freudian concept; otherwise, it will never be more than Microsoft's ill-fated TAY: a moron that is easily persuaded to become sexist and racist after only a dozen hours of conversation with users.

ANELLA: An associative network with emergent logic and learning properties

How do we go about that issue? We proceed along the way I proposed in the years 1987 to 1990. At the time I was a researcher in artificial intelligence within the framework of the British Telecom team to which I belonged as a fellow: the Connex project. I developed in those days an artificial intelligence piece of software, ANELLA (*Associative Network with Emergent Logical and Learning Abilities*), a very apt description for it, given by one of my colleagues, which would simulate emotions inside, that is to say that affect values would be attached to the elements of knowledge that this system contained.

Experiences in a human's life automatically generate emotions. Some are plainly pleasure related: when we eat something tasting good, the experience is more pleasurable than when we eat something nasty. Some satisfactions come to us in such and such way: we like to be complimented or praised and we don't like to be reprimanded, etc. If you've siphoned encyclopaedic knowledge into an AI and then wish it to be recalled in a relevant manner, the machine needs to know what is important in it: what is essential and what is accessory, what is most valued by some and what is not by some others.

What I had done with ANELLA was that a memory was built, but in the way a human being acquires it, i. e. there was a seed word, and that word was "mummy," and step by step, the child would connect other words to "mommy", like "daddy", like "brother", like "sister", like "milk", like "eat" and "drink", etc. To "mummy" first because the baby has needs and its immediate first needs are satisfied through its mother. You need to breath, you need to sleep, you need to eat, you need to drink, you need to pee, you need to poop, you have to sleep when you are tired. That's how we learn about life, and if we wish common sense knowledge to be acquired, that's how it comes to us. We don't sit in school with the teacher saying, "Here, this morning, I'm going to give a lesson in common sense knowledge": we acquire common sense knowledge essentially by interacting in everyday life with other human beings and trying to satisfy those needs of ours. Here lies the starting point.

Implicit in ANELLA was a learning dynamics which could be labelled "emergent" as each time a word appeared that was not known by the AI, it attempted to find a place for it within the network of its existing "knowledge space". In order to achieve that, it would state: "I don't know this word. Can I relate it to something I already know?" This is of course exactly what children do when they say: "What does it mean, 'preposterous' (or 'trigonometry', etc.)?" Parents know that in order to explain the problematic word they will need to connect it to some others that the child already knows. But the difference here between the machine and us is that with a human being, there are emotions, affect values, associated with words already stored, and their emotional tone will "contaminate" a new word that will find getting attached to them as the location it is longing for in knowledge space, giving it its "seed" affect value. A high affect value has become associated with the word "mommy" because of the high affect values linked to getting milk when you want it and not being happy when you don't get it, and whenever a new word will find "mommy" as an anchor in knowledge space, like "daddy" for instance, the affect value of "mommy" will act as a seed value for it, to be updated of course in later interactions.

This is the way to proceed, and here is what allowed that extremely simple AI piece of software, with a few tens of thousands of lines of programming only, to appear intelligent at a very low cost. At no time did ANELLA wish to utter the same sentences a second time because the affect values of the content words within it had been lowered inside ANELLA's memory automatically as soon as the sentences where those words were comprised had been uttered. As far as relevance was concerned, there was effectively a devaluation of what had just been said, not because it had failed to be interesting but because, having been uttered, there was a fair assumption that the user had fully grasped the message and the information content and didn't wish to hear it once more.

So there were two parallel principles of updating. With the first, the affect value of the words involved was dropping while ANELLA was talking and there was coming a moment when the AI was reaching a state of "I have nothing more to tell", just like with people speaking from the floor at a conference when, after having uttered a number of sentences, they stop at some point because they've said all they wanted to say. But at the same time, according to the second updating principle, when a conversation ended, the affect value of the words that had been used, those that had been put forward, was updated, either increased or decreased, according to the degree of appreciation, providing a new candidate starting point for later conversations.

We're going to try and give robots a history, and this will apply in particular to the robots that will replace us when we're no longer here, making sure that they produce a mimicry of human beings of a better quality than those they will have replaced. The recipe for doing so is robots whose life story is that of having acquired knowledge stepwise, a kind of knowledge supervised by "parents" and "teachers" who prevent them from developing an ethical system that would not be up to the task. These autonomous robots should have taken into account our mistakes and in particular all those mistakes we humans have made explaining why they are by then on their own, having taken our place, while we ourselves are gone.

So when you have produced an artificial intelligence of ANELLA's type, it won't occur as a problem that it becomes sexist or racist overnight because it's already shielded by the fact that it has mimicked in its learning process the build-up of a proper personality, however effectively short the process might have been in the case of an AI proceeding at a computer's speed. It goes without saying though that if an instance of ANELLA had been created and it turned out that its "parents" and its "teachers" were racists and misogynists, those traits would of course have been reproduced in it.

A Freudian implementable model of the human subject

As a logical entailment of what has just been asserted, an implementable model of the human subject will be now presented. This model derives from the works of Sigmund Freud and later psychoanalysts, with some additions due to the very purpose of reproducing a human subject as the product of a computer programme.

The reason why "human subject" is mentioned instead of "human being" is that central to the model aimed at is the notion that the "being" in question sees itself as a "subject", i.e. a person identified to a Self able to fight for itself, through, in particular, the use of the words pertaining to a language.

How to give robots common sense?

The debate on artificial intelligence is rendered opaque by the presupposition that reproducing in a robot what is proper to us human beings necessarily leads to the production of a machine with an artificial intelligence.

This is a naïve representation ignoring, on the one hand, that, similar in that respect to all other animals, the genus *Homo* has been endowed by nature with a single purpose, namely to reproduce itself, and this, whether it was entrusted to us by Heaven: "And God blessed them, and God said unto them, Be fruitful, and multiply, and replenish the earth" (Genesis 1: 28) or, from an atheistic perspective, by a self-replicating "selfish gene", and that most of our time is taken up with the ancillary tasks that enable us to fulfil the reproductive mission: breathing, drinking, eating, sleeping, protecting ourselves, disposing of waste, and, in the commodified world in which we live, "earning a living" to get the money to meet these needs.

The fact that we are "intelligent" has enabled us over the millennia to improve our security and comfort considerably, but intelligence is only incidentally and very occasionally involved in the tasks entailed by the needs to breathe, eat, drink, etc.

When we ask ourselves today, “How can we make a robot acquire an intelligence that is not specialised in such or such task (winning against an opponent in a game of Go, for example),” but an artificial “general” intelligence (“general” in the sense of being able to solve any problem), we forget that our intelligence is not essentially used to solve difficult problems such as “What is the level of inflation compatible with full employment? ”but to find a partner for our lovemaking, a good restaurant at lunchtime, a clean toilet when the need arises, etc.

How, then, can we endow an intelligent robot with “artificial general intelligence” (a question also called “common sense for the machine”) without simulating in this machine in a simple-minded manner the fact that it must eat, drink, breathe, make love, and sleep?

In fact, all the knowledge and more that this AI needs in the first place can be found in Wikipedia, and the rest it can learn as we do: by asking questions and finding out for itself, by experimenting.

But such knowledge would still only be words stuck together, and the robot must also have “emotional” intelligence; in other words, there must be “feeling” attached to the words it learns.

Libido comes first

Evidently, there is an anthropocentric bias in saying that “the species seeks to reproduce”, but the fact is that species do reproduce and that – when they are bisexed – they resort to this device of bringing together two sub-types in the population, namely males and females, and producing offspring from their conjunction, which ensures the replication of the species.

Whether some individuals end up not reproducing, or have no inclination to do so, is purely anecdotal as, on the whole, a sufficient number of them do, so as to keep the species living on.

As hardly needs to be reminded, human beings enjoy mating as 1) mating relieves a tension that keeps building up (the *libido* in Freudian parlance) and 2) the very act of mating is accompanied by feelings that, although they are of an aggressive nature (originating from within the brain centre for aggression), are nonetheless among the more pleasurable, if not the most pleasurable.

The sexual process implies the build-up of a tension within men and women, inducing them to get closer, i. e. an irritating feeling that vanishes once mating has taken place. While as soon as it has disappeared through a brutal gradient descent (as such is the operation from a physical point of view: that's how we model it), it will surge again from that point on, the tension getting restored little by little.

All other features of human behaviour derive directly or indirectly from the urge to reproduce. Day-to-day survival in particular is nothing but maintaining the setup for reproduction, i.e. the survival of the species. Throughout their lives, humans, in childhood, in adulthood when they are old enough to mate and reproduce, and afterwards, have to satisfy a certain number of urges: day-to-day survival encompasses breathing, drinking and eating, excreting, protecting oneself in various ways, sleeping, so as to rebuild our strength.

Human beings need in an initial stage to reach the age for fertile mating, then spend a number of years reproducing. In the whole period that precedes, i.e. childhood and adolescence, this is done without there being any real reproduction and we can consider that there is a so-called "latency period": a period during which the libido is only present under the embryonic form that Freud called "infantile sexuality". This is followed by the time of puberty when libido arises but intercourse still fails to be fertile. Then there is a period of fecundity when children are engendered through the mating of an adult woman and an adult man who are by then both fertile. Finally, the reproductive stage comes to an end: women cease to be fertile; men cease to be driven by libido and are therefore no longer attracted to women. When they have gone beyond the age for reproducing, their body decays little by little through ageing until they die due to the failure of one or a combination of organs.

Staying alive so as to reproduce implies satisfying some urges

Just as for reproducing, being breathless, being hungry, being thirsty, needing to go to the bathroom, or being sleepy are part of a process where discomfort grows until it is relieved in acts of pleasurable satisfaction such as a good meal, a good drink, a good pee, a good shit, or a good nap. When discomfort grows too big, one gets distracted, i.e. incapable of doing much apart from trying to relieve the urge.

The way we have to conceive things is that in order to allow reproduction to take place, the functions of eating, drinking, etc. must be ensured throughout life. And for each of these functions, we can represent this in the same way as for libido: there is a rise in hunger, then we eat, and there is satiety, i.e. the need falls to zero before getting restored. For instance, with eating, we can say that there are three moments: when we wake up, we soon start feeling hungry, we eat; then there is a period up to lunch when appetite rises again and we satisfy it; and when the evening comes, we are once more hungry and eat. Tiredness operates in a similar way: you wake up, then will gradually get tired during the day, you feel sleepy, you sleep, etc.

Instead of an animal aiming constantly at doing different things in a particular order, a human subject can thus be represented as attempting simply at ensuring *homeostasis:* "Homeostasis is the ability of a living organism to maintain certain internal characteristics of its body (temperature, concentration of substances, composition of interstitial and intracellular fluids, etc.) at a constant level" (Wikipedia), i. e. getting rid of the urge to mate, eat, drink, piss, poo, sleep when it becomes unbearable.

Plenty of our individual lives can be described satisfactorily in those basic terms of relieving those constantly renewed urges.

Delayed satisfaction and work

Once they've left their pristine abode, human beings have become accustomed to the delayed satisfaction of their urges.

We can add other characteristics. If you're in a society like I've known in Africa, the problem of eating and drinking is quite simple to solve because you can find stuff to eat and drink all around you quite easily: climb a coconut tree and cut a green nut where there's nourishing food and a refreshing drink; access to food and drink is immediate. There is no need either to look for a public toilet as you can go hide into the bush all around you and relieve yourself that way.

A constraint intervenes for human beings in a modern urban environment: the necessity of having money. You need indeed to pay for drinks apart from tap water and you need to pay for eating, and you need to pay for sleeping: it can be rent, or the full price of a home, it can be a hotel, it doesn't matter. There is thus an additional constraint on those urges that we've defined: the near necessity of working. You have to work a certain lapse of time and you know that working a number of hours will allow you to collect a certain sum of money by the end of the day and that amount of money can be used the following day to buy drinks, to buy food, to find a shelter where to sleep, and so on. So if we think of a particular person during a particular day, urges within her or his body mean she or he is subjected to certain constraints such that we can make her or his agenda for the day regarding not only basic needs but also the sexual tension rising from within: the libido. Mating has been concentrated on particular times in the day, in the week, and even in the year. Those are the constraints defining close to the full agenda for the day of a human subject. That particular observation may seem trite and trivial but it is proper to the psychoanalytical understanding of the human subject: Freudian metapsychology is sole in emphasising the peculiarity of the human fate.

It should also be noted that resting on the foundations of the social nature of humans as mammals, language has allowed cooperation between them to be further leveraged. Language has also contributed to adding much sophistication to the sexual parade observable in many other animals, allowing even the human subject to simply babble himself or herself into mating without the need for much gesturing.

The framework of an implementable model of the human subject has been thus provided in a nutshell. Its essential feature is that the human subject is acted by a double dynamics, one having an inner source, that of those urges which once satisfied keep building up again, and the other of an outer nature, the response that the natural environment offers to our attempts at relieving our urges. Remarkable in that respect is that the perception by ourself of the effects of our interaction with the world gets processed by us as information from an external source relative to interferences with the unfettered satisfaction of our urges in their constant process of renewed buildup. The very words we utter in particular are being processed by us as having either managed to satisfactorily satisfy the satiation of our urges or having on the contrary hindered it.

Memory as storage for procedural knowledge

Easing the smooth process of interaction between us and the world surrounding us is the incremental construction of a memory. Memory is constantly updated in response to two operating dynamics: one of external origin, induced by interactions with the world, and the other of internal origin, induced by our own impulses.

Memory offers us a blueprint for facilitated interactions with the environment. It is constructed both positively as promotion and negatively as inhibition from the respectively successful and unsuccessful ways we responded to the world opposing some resistance to our sheer exploitation of it.

In addition to a body, what equipment has a human being to help satisfy his various drives? Among other things, he needs decision-making principles to determine the order wherein to undertake the various operations that these satiations require. The information that allows our body to prioritise is stored in memory. To be able to determine an order of execution, that memory must be acted upon by a dynamics capable of evaluating the relevance of the range of possible actions at every moment, and of choosing among them the most relevant one: the one that should be taken in preference to the alternatives.

Memory has a double function: firstly, to be recalled at each moment in an uninterrupted evaluation of the present situation – memory offering us indications on what to do next from the information already stored; secondly, our perceptions of the events taking place at the present moment constitute the fund of new information, either relating to facts that we were previously unaware of and that we cease to ignore or relating to what is already known but which will allow us to complete, to update in whole or in part, prior knowledge.

Memory is therefore constantly the object of a double movement in opposite directions: stored memory, previously built, is called up to be put to good use in the context of the present moment, while the information contained in this present moment produces new memories which will be added to those already stored or will slightly modify their content, bringing them up to date, allowing us to refine their image, to nuance them, and ensuring the improvement of our performance during the recall of the memory which will take place when we find ourselves later in the same circumstances.

How is memory managed? This is where the psychoanalytical model comes into play. Three instances, which can be represented in a first sufficient approximation as real "agents", real actors, interact within a human subject according to Freud's second topographical model of the "mental personality", proposed by him in 1920. He had introduced his first topographical model in 1895, wherein there were three zones rather than actual agents: the unconscious, the preconscious, and the conscious.

Three agents: the Id, the Ego, and the Super-Ego

In Freud's second topographical model of the "mental personality", playing the role of an infrastructure, lies the least accessible part of the unconscious called the "Id", the term Freud uses for it in his theoretical model. Then there is the "Ego", which is conscious for its most part: the "Self" we assume we are in essence and suppose is the master in full control of our willpower. Finally, there is the largely unconscious but partially consciously accessible "Super-Ego", an instance which was traditionally called the "voice of conscience" and, even earlier on, in our culture, our "guardian angel".

The Id

The essential functioning of the machine that is the human being and its maintenance is ensured by the Id: a handyman. Not only does the Id watch over our reflexes, but it also takes over the entire machine as soon as our attention is captured elsewhere. When we sort out the children's bickering in the back seat, it is the Id that ensures that the car does not roll over into the ditch.

The Id, if we go back two millennia in the history of our civilisation, is what Saint Paul (Paul of Tarsus) called "the flesh": a second will, distinct from that which he designates as the "I", that expresses itself also under his name, and which is antagonistic to that "I" which equates with what psychoanalysis today calls the "Ego". Paul wrote thus in one of his epistles:

18 For I know that in me (that is, in my flesh,) dwelleth no good thing: for to will is present with me; but how to perform that which is good I find not.

19 For the good that I would I do not: but the evil which I would not, that I do.

20 Now if I do that I would not, it is no more I that do it, but sin that dwelleth in me.

Epistle to the Romans 7 (*King James Version*)

Freud wrote: "To the oldest of these psychical provinces or agencies we give the name of id. It contains everything that is inherited, that is present at birth, that is laid down in the constitution – above all, therefore, the instincts, which originate from the somatic organisation and which find a first psychical expression here in forms unknown to us" (Freud [1938] 1940: 2).

By definition, of course, all processes taking place without appearing to consciousness remain unconscious, and for this reason we call them "automatic" as we cannot deny that they go along their course when we're "thinking about something else", when "our mind is drifting elsewhere", etc. For example, I am not paying any attention to the fact that for a while already I need to pee, but my body, under the direction of the Id, is locating itself and looking for a place where I can go and satisfy my urge. This initiation of the search is not deliberate: it is unconscious; I don't think about it. It will happen that in other circumstances, having ignored the passing of time for too long, I will suddenly say to myself: "Well, now I really have to find a place to pee as otherwise I'll start urinating on myself." At this point in time, the conscious "I" has taken over.

The Ego

This subjective sense of full presence in the world that is consciousness emerges at the crossroads where memories of situations similar to those we are experiencing intersect with the memories we are creating in "real time". Each of these memories – already registered or in the process of being registered – carries with it a mood, an affective climate, which is its own: that of the past when it was first registered and that of today in its new registration.

Day-to-day survival requires only marginally consciousness where the Ego is in our representation in the driver's seat. Most of the process and maintenance is provided by the Id in Freudian parlance: an all-purpose caretaker. The Ego at the centre of consciousness is, however, summoned in deliberate planning and implementation progressing from carefully planned step to step, constituting so many intermediate goals.

If we reason in terms of implementation, then the Id requires an original type of representation, such as the one I had turned to in the programming of my own ANELLA AI piece of software. The dynamics of ANELLA consists of paths being followed on a directed and weighted graph representing stored mnesic traces and constituting as a whole a model of an individual's memory, organised as the nature of its support imposes: a natural neural network composed of interconnected neurons – whereof artificial neural networks such as those used by current *deep learning* systems offer but a very simplified approximation.

And the difference between the unconscious and the conscious Ego is that if I'm "thinking about something else" or if I don't think about it at all, I will unconsciously go to the toilet and pee, without having consciously formulated the intention to do so as well as the will to carry out my intention, followed by its implementation. In other words, the Id will have taken care of all of this, from the latent, implicit intention to its realisation. Of course, if I procrastinate and it comes to my attention, i.e., is displayed in the window of the conscious Ego, from now on I must deliberately perform certain actions, because from now on I must "really" pee, then the Ego engages. And what the Ego can do, which exceeds the capacities of the unconscious, is to plan in a deliberate way, to say to myself that I must now satisfy the urge, for example, in the next 5 minutes, and to do it, possibly in stages, that is to say, by giving myself intermediate goals, stages of which one can consciously enumerate the order wherein they must be done and then perform them in that order.

It is safe to say that as far as the functions of the Ego are concerned, AI in its current state of research and development has been able to formulate them in the form of mostly familiar algorithms.

So there is an instance stemming from the memory whose behaviour is automatic, and that is the Id. And there is another instance that can call upon memory to deliberately plan operations, and that is the Ego. And there is a third instance in Freud's second topographical model of the "mental personality", which is part of the Freudian setup of the human subject, and that is the "Super-Ego".

The Super-Ego

The Super-Ego is grafted onto the Id; it embodies a part of knowledge under the shape of behavioural rules that have not been acquired through personal experience but by a shortcut as the experience of one's parents and of the surrounding culture as a whole. They are rules to follow or things to do that parents have promoted, that teachers have recommended; they are views of mentors or have been discovered by oneself in authors one admires. The set amounts to what French sociologist Emile Durkheim (1858–1917) called the "interiorised social". The Super-Ego's set of rules are hierarchical, with some of those having ascendancy over others.

The Super-Ego may, however, have been endowed with either inefficient or impractical tyrannical rules of thumb, encapsulating the errors of our forebears over the ages. While the Id operates "intuitively", that is to say, by means of the non-linear effects of a directed (natural) neural network, the Super-Ego is more of the nature of an expert-system applying a hierarchical set of rules to the raw outputs produced by the Id, acting as a filter that operates on these raw outputs to make them polished ("policed").

Most of the time, the rules that the Super-Ego is made of do not emerge to consciousness but this doesn't prevent them from imposing themselves by bending to their norms the instinctive behaviour which is the realm of the Id. The Super-Ego manages to impose its rule but so to say back-stage, interfering with the way that both the Id and the Ego operate separately and in the dialogue between them. Once those implicit rules have reached consciousness, they may of course be explicitly stated.

It is possible to come up with a very economical and clear representation of what would be the raw output of a model of the Id as a directed and weighted graph when it passes through the filter which is the set of rules constituting an expert-system modelling of the Super-Ego. But when the processing operates through the mechanism we call "intuition", no rule is applied: in that case, it is the activation of the neural network as it is in itself, i. e. constituting a whole. That is what we call "intuition": it is reasoning taking place within ourself but according to a mechanism to which we have no access and which remains opaque to us: we do not know exactly what happens within. We say, without being able to explain it further: "it is of the order of the unconscious".

Mimicking a human subject is not the same as making an intelligent robot

It is hardly necessary stressing how different a starting point there is between a machine such as a robot and a human subject, with all urges for reproduction and survival being absent from a machine. A sentient robot would need those to be animated by a proper simulated affect dynamics. To test the views expressed here, a male and a female robot should be created having over the different hours in the day urges in a lower or higher degree to be relieved, determining their behaviour and interest in each other. In simulation mode the interplay could then be observed between several instances of such robots.

It is unlikely that, as part of an AI overall project, we would ever feel the need to replicate a human being with the entirety of its urges linked to being a creature geared at reproducing itself. Indeed as the label aptly implies, "artificial intelligence" is focused on a single feature of the human complex: its intelligence.

The issue of the relationship between intelligent machines and us is thus per definition dramatically restricted to a single dimension of what makes us human.

The difficulty with intelligence is that we essentially recognise it when we see it and are not particularly good at defining it precisely. And that difficulty is considerably enhanced when we're talking of super-intelligent machines to come, i.e. being better than we personally are at being intelligent in the intuitive way we assign to that notion.

What do we expect from an AI?

It is at that juncture that Freudian metapsychology has a crucial role to play: not at refining our definition of AI but at understanding in a much clearer way what it is we expect from so-called "intelligent machines", having in mind the delicate interweaving and interaction between the Id, the Ego, and the Super-Ego that constitute us. What is it we want to tell machines of our goals with them, and what can they expect in return from us? Does it require that we develop – on top of programming – a specific language for talking with machines? It could very well be the case; Stephen Wolfram for one believes it to be the case:

> "... we don't recognise it as 'intelligence' unless it's aligned with human goals and purposes [...] we're going to have to define goals for the AIs, then let them figure out how best to achieve those goals. [...] the real challenge is to find a way to describe goals. [...] we need to tell them what we generally want them to do. We need to have a contract with them. Or maybe

> we need to have a constitution for them. And it'll be written in some kind of symbolic discourse language, that both allows us humans to express what we want, and is executable by the AIs. [...] In a sense the constitution is an attempt to sculpt what can happen in the world and what can't. But computational irreducibility says that there will be an unbounded collection of cases to consider" (Wolfram [2017] 2020: 556, 561–562).

Plan C: a world populated by autonomous robots, from which we will have disappeared

Returning now briefly as a matter of conclusion to Isaac Asimov, the father of the "Three Laws of Robotics", he had to say the following fateful words:

> I wish I could say that I am optimistic about the human race, but I fear that we are too stupid and short-sighted. And I wonder if we will ever open our eyes to the world around us before we destroy ourselves.
>
> [...] when the time comes when robots, wishfully, become sufficiently intelligent to replace us, I think they should. We have had many cases in the course of human evolution and the vast evolution of life before that where one species replaced another because the replacing species was in one way or another more efficient than the species replaced. I don't think that homo sapiens possesses any divine right to the top rank. If there is something better that we are than let it take the top rank. As a matter of fact, my feeling is that we are doing such a miserable job in preserving the Earth and its lifeforms that I can't help feeling that the soonest we are replaced, the better for all other forms of life. (Asimov 2022).

Indeed it is probably much more feasible to work on developing machines, robots, that will replace us entirely than to try and save the human race in the current context of its presence on our planet: its having trespassed Earth's carrying capacity for such a voracious and ill-behaved species. This is a view I advocated back in 2016 in *Le dernier qui s'en va éteint la lumière* ("The Last One to Leave Turns Out the Light").

So, thinking of Plan C of humans replaced by robots, when I claim that it is the most feasible project compared to other tasks like saving humankind as Plan A, I don't mean to say that it is feasible in the sense that the chances are enormous that the mission can be completed. I mean that, in a comparative perspective between other tasks and this one, for example, as human beings settling on other planets and living there autonomously as Plan B, compared to that, the task of creating autonomous robots that would reproduce is, in my opinion, the easiest one of the three to achieve because I personally don't see any major technical obstacles remaining to its success, only time needed for normal research and development, that is, if artificial intelligence is wise enough to choose as a blueprint for the human being to be emulated the one that Sigmund Freud displayed with his "metapsychology" of psychoanalytical inspiration, a masterpiece of scientific achievement in an environment where experimental setups were – and remain – nearly impossible to come up with.

References

Asimov, Isaac, *I, Robot*, [1950] London: Grafton Books 1986

Asimov, Isaac, *The Rest of the Robots*, [1964] London: Panther Science Fiction 1968

Isaac Asimov, l'étrange testament du père des robots, a documentary by Mathias Théry (Fr., 2020, 55 min), October 2022

Freud, Sigmund, *An Outline Of Psycho-Analysis*, [1938] London: The Hogarth Press 1940

Jorion, Paul, *Principes des systèmes intelligents*, Paris: Masson 1989; reprint Broissieux: éditions du Croquant 2012

Jorion, Paul, *Le dernier qui s'en va éteint la lumière*, Paris: Fayard 2016

Moor, James H., "Four Kinds of Ethical Robots", *Philosophy Now* 72 2009: 12–14.

Murphy, Robin R. and David D. Woods, "Beyond Asimov: The Three Laws of Responsible Robotics", Intelligent Systems, IEEE 24(4), September 2009: 14–20

Wolfram, Stephen, *"A New kind of Science: A 15-Year View"*, May 16, 2017, in *A Project to Find the Fundamental Theory of Physics*, Wolfram Media 2020

Hub Zwart

Psychoanalysis and Artificial Intelligence: Discontent, disruptive algorithms, and desire

Abstract: Starting with Freud's concept of the psychic machine (Entwurf; *Interpretation of Dreams*; *Beyond the Pleasure Principle*) I will discuss Lacan's effort to elaborate this view with the help of 20th-century research areas (computer science, linguistics, cybernetics, molecular biology, etc.), resulting in the famous theorem that the unconscious is structured as a language. Subsequently, two closely related questions will be addressed, resulting from a mutual encounter between psychoanalysis and AI, namely: How can psychoanalysis contribute to coming to terms with AI and to what extent does AI allow us to update psychoanalytic theories of the unconscious?

Keywords: computer science, Lacan, cybernetics, linguistic

Introduction ("Eurydice deux fois perdue")

This contribution aims to address two complementary questions: (a) how can psychoanalysis allow us to come to terms with current developments in artificial intelligence, in society as well as in science, and (b) how can current developments in artificial intelligence allow us to redefine the epistemic specificity of psychoanalysis? In addressing these questions, I will build on the work of Jacques Lacan, notably his seminars: Lacan's laboratory, as it were, enacting a mutual exposure exercise. On the one hand, these seminars entailed a return to (i. e., a close *rereading* of) the work and experience of Sigmund Freud, so as to determine the *specificity* of psychoanalysis which, according to Lacan, had become obfuscated by recent developments such as ego psychology (aiming to bring psychoanalysis in accordance with the dominant neo-liberal ideology of modern technocratic capitalism). On the other hand, Lacan's seminars contain multiple references to emerging developments in twentieth-century research fields (e. g., linguistics, ethology, computer science, cybernetics, biomolecular genetics, etc.). In all these confrontations, Lacan not only developed a psychoanalytic perspective on contemporary science and technoscience (Zwart 2022a) but also used the insights and vocabularies of these research developments to elucidate and rephrase the epistemic specificity of psychoanalysis itself as a revelatory truth event.

https://doi.org/10.1515/9783111007564-003

To phrase it dialectically: at first glance, the revelatory and confrontational insights entailed in Freud's writings seemed to be *negated* by twentieth-century technoscientific developments. While ego psychology tried to adapt psychoanalysis to the liberal ideology of autonomy, competitiveness, and performance, thereby obfuscating its critical potential, technoscientific developments tended to see psychoanalysis as unscientific and outdated, as a mystification or aberration, and as something of the past. Lacan aimed to *negate or supersede this negation* by demonstrating how precisely these recent scientific developments allow us to emphasise the validity, urgency, and relevance of a psychoanalytic perspective on contemporary science, technology, and political culture.

My contribution aims to recall Lacan's dialectical approach, focusing on artificial intelligence as an emerging research field. First of all, I will zoom in on what Lacan considers the basic epistemic insight of psychoanalysis, namely the disparity between subject (psyche) and object (external reality) – a chronic discordance which human civilisation (notably its technological dimension) tries to overcome. Whereas other species dwell in a natural environment, to which they have more or less become adapted in the course of evolution, technology created an artificial environment, a world of language, concepts, rules, institutions, machines, and artifacts, referred to by Lacan as the symbolic order. Science entails a symbolisation of the real with the help of mathematical and technoscientific symbols, amounting to a disruptive reorganisation of natural environments. And artificial intelligence must be considered a decisive final stage in this development. Instead of gratifying our needs, however, the symbolic order amplifies our sense of deficiency and intensifies human desire, and this applies to AI as well.

Against this backdrop Lacan notably emphasises, in an anticipatory manner, the importance of gadgets, in which technoscience currently objectifies itself – high-tech entities of relatively small size that are entirely forged by science and are transforming the symbolic order. Again, although these gadgets purport to fulfil our wishes and enhance our lives, their overwhelming presence actually represents an existential challenge, giving rise to anxiety and discontent.

In addition, however, Lacan points out that artificial intelligence also allows us to deepen our understanding of the specificity of basic psychoanalytic concepts (such as the unconscious, Freud's most decisive discovery) by making them more precise. In ego psychology, the ego aims to safeguard its autonomy vis-à-vis the drives that are supposedly contained within the Id. This conceptualisation builds on Freud himself who, in one of his final manuscripts, entitled *An Outline of Psychoanalysis* ("Abriss der Psychoanalyse"), explained that the psychic apparatus consists of three provinces or agencies: the ego, the super-ego, and the id. The latter is the oldest of the three and contains *everything that is inherited*, everything that is constitutionally present at birth, and above all, the instincts ("Triebe"), which originate from the somatic organisation and which find their first psychical expression here.[1] This definition, emphasising basic instincts originating from the body, seems fairly open to a biological interpretation of the unconscious, presenting the Id as a kind of inner animal,[2] to be domesticated by upbringing, society, and culture, in collaboration with the ego (and its defence mechanisms) and the super-ego (as the internalisation of societal restrictions and demands). The phrase "everything that is inherited" seems to suggest that the Id might be regarded as the sum of our (unconscious) genetic predispositions.

Lacan vehemently rejected a biologistic reading of Freud, considering it a misinterpretation that obfuscated the specificity of psychoanalysis and its radically critical stance towards the dominant liberal or neo-liberal ideology of contemporary Western civilisation. For Lacan, the unconscious should not be seen as the wild animal within, although this interpretation is reinforced by the English translation of *drive* (*Trieb*) as *instinct*. Building on a meticulous rereading of Freud's oeuvre, Lacan persistently argues that such an interpretation misrepresents the genuine and unprecedented significance of Freud's key discovery, the discovery of the unconscious. As if Eurydice (temporarily brought to life by Orpheus-Freud) is allowed to disappear, to slip away again.[3] To keep the authentic Freudian concept alive, to save it from these misinterpretations, it must be drastically reframed.

The conceptual problem is caused by a chronic ambivalence that runs through the work of Freud himself. Although he was trained as a neurophysiologist of the nineteenth-century positivistic school, his discovery of the unconscious entailed an epistemological rupture: a fundamental departure from his earlier (scientific) work. Still, his intellectual upbringing resulted in a kind of wavering, with Freud on some occasions stressing the uniqueness of psychoanalysis (as an endeavour *sui generis*) while on other occasions cherishing the hope (or even the expectation) that one day its basic concepts would be reinterpreted in biological terms and re-embedded in biology: a chronic wavering between positivistic and post-positivistic understandings of psychic life (Ellenberger 1970). According to Lacan, this ambiguity can be addressed by reframing Freudian psychoanalysis with the help

1 "Die älteste dieser psychischen Provinzen oder Instanzen nennen wir das *Es*; sein Inhalt ist alles, was ererbt, bei Geburt mitgebracht, konstitutionell festgelegt ist, vor allem also die aus der Körperorganisation stammende Triebe, die hier einen ersten ... psychischen Ausdruck finden... Dieser älteste Teil des psychischen Apparates bleibt durchs ganze Leben der Wichtigste..." (1938/1941b, p. 67/68).

2 The unconscious has often been regarded as our "inheritance from the animal world" (Sulloway 1979/1992, p. 4).

3 "Eurydice deux fois perdue" (1964/1973, p. 33)

of twentieth-century research fields such as linguistics, molecular genetics, and cybernetics. These research fields allowed Lacan to redefine the unconscious as being structured like a language, a code, a chain of signifiers, comparable to some extent to a text, a computer code, or DNA (Zwart 2013).

The orphic aspiration to retrieve the radical specificity of psychoanalysis is not a one-time event, but rather an interminable assignment. While technoscientific developments continue to see psychoanalysis as something of the past – as an outdated view that must be or de facto *has been* replaced – Lacan rather argues that these technoscientific developments not only force us but also *allow us* to redefine the specificity of psychoanalysis as a research field *sui generis*, a critical alternative to mainstream scientific and ideological endeavours.

To clarify what this amounts to, I have opted for a case study approach, by submitting a recent hyper-popular science bestseller, namely Yuval Harari's *Homo Deus* (2015/2016), to a critical Lacanian reading. This bestseller aims to bring together current technoscientific insights generated by computer science and related research fields into a technoscientific anthropology and worldview. Harari's book exemplifies the emerging synthesis between neoliberalism and computer technology. Basically, building on artificial intelligence and computer science, human beings are redefined as "aggregates of algorithms". At first glance, a Lacanian reader will notice some level of concordance between Harari's interpretation and the views of Lacan. To some extent, both the Lacanian unconscious and the symbolic order can indeed be redefined as "aggregates of algorithms". At the same time, precisely Harari's key concept (the algorithm) allows us to specify how psychoanalysis radically diverts from a technoscientific neo-liberal worldview. To frame it in dialectical terms, while *Homo Deus* purports to be a final *negation* of psychoanalysis – aspiring to eliminate the last remnants of "Freudian psychology" from the vocabularies of technoscience once and for all (p. 136) – a Lacanian reading of this bestseller entails a *negation of this negation* by arguing that Harari's basic concept (algorithms) allows us to demonstrate why a psychoanalytic understanding of civilisation and the human psyche is both radically different and urgently called for – more than ever.

Disparity

According to Jacques Lacan, the decisive originality of psychoanalysis (the *epistemic rupture* initiated by Sigmund Freud around 1900) was to denounce the "pastoral" idea of an original harmonious relationship between subject (psyche) and object (external reality) which we should somehow try to re-establish and restore (Lacan 1959–1960/1986, p. 107). Instead, Lacan argues, building on Freud (1920/

1940), that a chronic disparity between both poles should be our point of departure. As I have argued elsewhere (Zwart 2013), an intriguing concordance can be discerned between the discoveries of Freud and the rediscovery (also in 1900) of the work of Gregor Mendel, pointing at the existence of a biological algorithmic code or program, thereby revealing a structural tension between the instructions contained in this program and the demands and restrictions coming from the external environment. An organism can only thrive to the extent that this tension can be superseded.

This tension between the unconscious program and the external real is amplified in the case of human beings. The initial encounter of human beings with the threatening real is a traumatic experience, exemplified by the trauma of birth. As Slavoj Žižek (2016/2019, p. 157) points out, Immanuel Kant already interpreted the screams ("Geschrei") produced by a child at birth as a symptom of indignation in response to the experience that human autonomy is significantly hampered by the insufficiency and vulnerability of our bodies vis-à-vis the real. For psychoanalysis, the birth trauma emphasises the maladaptation of human organisms to their natural (primal) environment. Since our struggle for survival cannot solely rely on our program, we develop additional systems (technology and civilisation) to mitigate the exposure to the real. Yet, as an artificial environment, these additional systems prove a highly demanding environment as well, confronting us with excessive or even impossible expectations.

The traumatic experiencing of a threatening real is not something which only applies to humans shortly after birth. Rather, it is a basic constituent of the human condition. The birth trauma indicates that we experience our real body as a fragmented body: disrupted from our intimate connection with the feeding womb and breast. All subsequent efforts to supersede this experience of disruption remain questionable and fragile. Our sense of identity and individuality remains vulnerable vis-à-vis the threatening real. According to Freud (1920/1940), an important objective of human civilisation in general, and of its technological dimension in particular, is to immunise our body and psyche against the real and to safeguard our fragile integrity from disruptive external intrusions. Contrary to most other mammals, human beings enter prematurely into the world, and their existence remains tainted by negation and lack (e.g., the *absence* of fur, claws, etc.). This is especially noticeable in neonates (e.g., their *inability* to move and walk), so that, with the help of technology, additional immunisation devices (a cradle, a baby carrier, a home, etc.) must be installed to compensate for these lacks which are threatening human existence with negation and elimination from the very outset. Dialectically speaking, psychoanalysis sees technology as the *negation of this negation*, as an effort to supersede the disparity between program and environment, the primal experience of lack (Zwart 2022). Rather than being "open" to externality and otherness, our existential challenge as humans initially comes down to averting, neutralising, and, to a limited extent, incorporating this threatening avalanche of external input.

These views were already proposed by Freud during the early years of psychoanalysis in the 1890s, in his letters to Wilhelm Fliess as well as in a manuscript known as the *Entwurf*, published posthumously (Freud 1950). In these documents, Freud describes the human psyche as a "machine" (p. 139), an "apparatus" (p. 270) consisting of various "systems", wherein energy quanta circulate, designed to attenuate excessive stimulation and excitation. Indeed, "I am a machine" (1950, p. 271; cf. Zwart 1995) and the main function of this machine (the neural apparatus) is to act as a screen ("Quantitätsschirm") to contain the influx of potentially disruptive energy quantities, entering the system from outside (p. 390). The psychic apparatus acts as a filter which allows only small quotients of external energy quantities to affect the psychic system (p. 394). Thus, the main task of the psychic machine is to protect the system from intrusion by disruptively large quantities of input. A plethora of technologies developed by humans can be considered extensions and externalisations of these psychic mechanisms, fostering immunity by strengthening our capacity to safeguard our psychic and physical integrity. Sense organs, either natural or artificial, function like antennae, allowing only small samples of the raw (and potentially overwhelming) real to be processed. Therefore, Lacan distinguishes "reality" from "the real". What we experience as reality is the outcome of an intricate and complicated process. Human reality is drastically filtered, processed, and construed.

At first glance, the role of the "reality principle" seems to be to enhance the ego's ability to defer immediate gratification (pleasure). Yet, on closer inspection, the role of the reality principle first and foremost is to *shield* the ego, by forfending traumatic *confrontations* with raw externality. The primary role of the reality principle is not to *expose* the subject to the inexorable real, but rather to allow *carefully selected bits of reality* into the system so that these samples ("raw quantities") can be adequately processed, and reality becomes liveable for the subject. In short, whereas traditional philosophy emphasises world-openness and intentionality as starting point for human understanding, psychoanalysis rather emphasises the epistemic role of resistance as a mechanism of defence (Zwart 2019a).

This line of thinking is taken up by Freud many years later, in *Beyond the Pleasure Principle* (Freud 1920/1940). The pivotal role of resistance, Freud argues, is underscored by human anatomy. We are covered by protective skin (which is subsequently covered with an additional protective layer known as clothes), while our sense organs are miniature apertures whose primary purpose is to provide protection against overstimulation (*Reizschutz*). Rather than being open to the world, our bodies protect and immunise us from the threatening Real. This tendency of living organisms to insulate themselves from the outside world already applies to micro-organisms, coaxed inside their cell membranes. Our vulnerable bodies protect themselves against overstimulation, but this applies to the human psyche as well. Protection against external stimuli is a life task at least as important as sensitivity and receptivity (Freud 1920/1940, p. 27). As indicated, our sense organs are like little antennae that select small samples of exteriority, allowing us to assess minute quantities of input. Our primary objective is to safeguard our psychic integrity from intrusive traumas.

Freud elucidates the topology of the human psyche by comparing it with the anatomy of the human eye. Darkness is the default, and the eye is basically a camera obscura, while pupil and cornea allow only small samples of diffracted light to enter the eye and reach the retina. Raw light is meticulously filtered and processed.

We may see a laboratory as an extension or externalisation of the human eye: as a space where everything (light, air, temperature, etc.) is meticulously conditioned and controlled, safeguarded from external disturbances so that only carefully selected samples of reality are admitted and subjected to analysis, with the help of precision contrivances. What the example of the laboratory also indicates, however, is that the scope of our vision (of our sensitivity) can be significantly broadened with the help of artificial extensions, namely artificial sense organs and electronic gadgets, so that humans gradually evolve into the "prosthetic superhumans" (Freud 1930/1948). This means that, after immunisation and selection, the next challenge is the threat of overcompensation. Paradoxically perhaps, while initially designed as immunisation devices, technologies eventually tend to evolve into sources of information in their own right, bombarding us with input. In the global high-tech environments of today, humans are exposed to technologically mediated overstimulation (information overload). While laboratories may initially be considered materialisations of Freud's concept of the psyche (operating as a highly selective immunisation device, a drastically simplified version of the external world), the currently emerging global networks of laboratories are confronting us with informational overabundance (data litter). Knowledge scarcity has definitely given way to gargantuan data collections. The whole world is becoming a global laboratory.

Let this suffice as a starting point for outlining a psychoanalytic approach to understanding contemporary technology. I will now zoom in on the added value of Lacan's assessments of contemporary technoscience against the backdrop of his rereading of (return to) Freud.

1953

1953 is an important year for science in general, but for psychoanalysis in particular (Zwart 2022). It is the year of the discovery of the biochemical structure of DNA by James Watson and Francis Crick, but it is also the year in which Jacques Lacan inaugurated his famous *Seminars* (Lacan 1953–1954/1975). In these seminars, technoscientific breakthroughs (such as the discovery by Watson and Crick) are assessed from a psychoanalytic perspective but, at the same time, used to explore and flesh out the specificity of psychoanalysis itself. As indicated, these seminars operated as Lacan's laboratory, designed to enact an intensive mutual exposure between the writings of Freud on the one hand and the vicissitudes of twentieth-century technoscientific research on the other. Lacan's explicit objective was a *return* (a systematic rereading) of the writings of Sigmund Freud, but rather than opting for an orthodox "author studies" approach, his effort was to retrieve the revelatory truth of psychoanalysis by mutually confronting Freud's discoveries with the ground-breaking vicissitudes of technoscience.

Freud's revelatory insights had been obfuscated by post-war developments such as ego psychology, Lacan argued, and his objective now was to *negate this negation* of Freud's revolutionary insights. The provocative originality of psychoanalysis had been obfuscated by an ideological misreading of his work, bent on "strengthening the ego", which *de facto* came down to forcing the ego to adapt to its socio-cultural environment, thereby negating what Lacan saw as Freud's decisive, non-conformist, revelatory truth. Therefore, the aim of his seminars was to supersede this obfuscation (to negate this negation) – so that his seminars amount to an exercise in retrieval, restoring Freud's original insights, albeit at an advanced level of sophistication and comprehension.

Therefore, rather than merely *reading* Freud, Lacan opted for triangulation, by confronting Freud's oeuvre with important developments in twentieth-century research fields unknown to Freud himself. Thus, Lacan systematically reframed Freudian conceptions with the help of terms and insights adopted from research fields such as linguistics, cybernetics, ethology, informatics, molecular biology, and so on. Freud himself already anticipated the need for such an endeavour. He wrote extensively, for instance, about how language studies research into the antithetical meaning of primal words (Freud 1910/1943) and anthropological research into aboriginal societies could help us to come to terms with contemporary neurosis (Freud 1913/1940). And he also wrote about the psyche as a machine, as we have seen, exploring pathways of research that would later be taken up by cybernetics and neuroscience. Yet, where science as such was concerned, Freud remained very much oriented on his personal experience as a participant in research fields of the late nineteenth century, such as neurophysiology, in which he himself had been trained, while he was virtually unaware of revolutionary developments which transformed multiple areas of research from 1900 onwards (linguistic, genetics, quantum physics, and so on). By mutually exposing Freud's writings to contemporary research fields, Lacan intended to retrieve and reaffirm the provocative originality of psychoanalysis.

This inevitably implied that Lacan's own thinking increasingly moved beyond the nomenclature and parameters designed by Freud. Notwithstanding Freud's emphasis on disparity, especially in unpublished fragments such as a short unfinished essay on *Ich-Spaltung* (1940/1941a), avidly discussed by Lacan, for many readers Freud's starting point seems to be a self-contained ego who gradually opens up to reality: a developmental trajectory which moves from inside to outside as it were so that an ego exclusively focused on libidinal drives gradually learns to redirect part of its energy in external things (Aydin 2021). Lacan radically reverses this thesis: the ego comes into being *via exposure to otherness*. Thus, Lacan radically reinterprets the famous Freudian adage *Wo Es war soll Ich werden:* Where it (or Id) was, I (ego) shall come into being. Freud presents this as a project of psychic reclamation so that the wild unconscious drives become cultivated and transformed into a polder landscape, a metaphor which seems to suggest a gradual strengthening of the ego. Lacan's starting point is different. Otherness precedes the self. Individuation requires a symbolic order.

The Symbolic

For Lacan, the premodern, Aristotelian-medieval cosmos was basically a "phantasy" (Fink 2004, p. 148), revolving around the idea of a pre-established harmony between *world* (macro-cosmos) and *psyche* (micro-cosmos). Via quantification and formalisation, technoscience disclosed a universe in which human existence is radically de-centred. This "narcissistic offence" (Freud 1917/1947) gave rise to a split and marginalised subject. And yet, there is something special about humans because, rather than in a natural Umwelt, we dwell in a "symbolic order": an artificial environment consisting of networks of signifiers (prohibitions, regulations, written and verbal instructions, textual messages, quantitative information, and so on).

Humans are speaking beings, called upon by language, by the commanding word, the discourse of the Other: the symbolic order which, for humans, is always already there. What is unique about humans, according to Lacan, is neither their intelligence nor their convoluted brains, but first and foremost, their openness to language. If IQ would be the decisive issue, human intelligence (as the outcome of

Darwinian evolution) would have been up to its tasks, allowing us to smoothly adapt ourselves to our environment (Lacan 1963/2005, p. 72). In humans, however, we rather see a chronic *failure* to adapt, a discord between desire and environment. It is precisely here, in human discontent, that language intervenes. Language has a disruptive impact on human existence. We are *speaking* animals, liberated from nature, but burdened by language, even sick with language (Lacan 1974/2005, p. 90, p. 93; Cf. 1961–1962, p. 42). And at the same time, language offers us alternative venues to articulate and explore our desire.

Due to language and other dimensions of human culture building on it (including technoscience), a decisive rupture separates human existence from the natural (pre-symbolic) mammalian world. According to Lacan, without language humans would be happy animals thriving in a natural Umwelt, where visual cues (described by ethologists as *stimulus* or *Gestalt*) would unleash pre-established physiological mechanisms (1953/2005, p. 20), pre-programmed behavioural responses (fight, flight, freeze, arousal, etc.). As animals, humans would dwell in an ambiance of visual gestalt-like stimuli, referred to by Lacan as "the imaginary": basic sets of images, and the repertoire of typical responses triggered by them. But the human world is replete with and disrupted by "the symbolic": norms and expectations, numerical and linguistic information, giving rise to a supra-personal "symbolic order". It is because of this symbolic order that science can exist, allowing us to come to terms with the Real with the help of a terminological grid of technical terms and other symbolic ingredients (numbers, formulae, measurements, mathematical and chemical symbols, equations, computer programs and the like).

For Lacan, scientific research tends towards "symbolisation", transforming the geosphere and biosphere with the help of "characters". In ancient Greek, στοιχεῖα (elements) refers to elementary building blocks (of reality or knowledge) but also to characters of the alphabet (letters and numbers), and this applies to modern science as well. According to Lacan, science is the systematic effort to disclose the basic constituents of nature with the help of symbols: Arabic numbers, alphabetic letters, mathematical symbols, chemical formulae, and so on. These numerical or letter-like (typographical) symbols are the "elements", the symbolic "atoms" by means of which science operates (1960/2005 p. 23, p. 50). Thus, whereas the pre-scientific world of everyday experience continues to rely to a significant extent on images (visible entities, world views, body images, self-images, metaphors, anthropomorphic interpretations, and the like), technoscience develops contrivances (measuring instruments, experimental equipment, etc.) which replace these imaginary, gestalt-like items with standardised terms, numbers, digital data, and equations. Molecular genetics, for instance, aims to *see through* the living organism (the visible gestalt) in order to *read* the symbols (the "characters") within – the genotype in the literal sense of "type" (Zwart 2016). Insofar as science produces images, they are highly technological, such as crystallographic X-ray pictures of DNA: visualised quantifications (Lacan 1961–1962, p. 42). The symbolisation process gives rise to a terminological grid of signifiers and quantitative numerical data. This means that the scientific universe is a radically "inhuman" world (1960/2005, p. 49). Science abstains from anthropomorphism (the tendency to interpret the world from a decidedly *human* viewpoint, p. 50).

Through symbolisation, the organic world of organisms (the biosphere) becomes incorporated into a symbolic ecosystem. Reality is "obliterated" by the intervention of the symbolic (Lacan 1966, p. 654), by the advent of numbers and probabilistic thinking, re-ordering the world in terms of the digital logic of presence or absence, affirmation or negation, as a method for structuring the Real (p. 655, p. 682). Etymologically speaking, "digit" is derived from *digitus* ("finger"): the *index* ("forefinger") indicating presence or absence. Thus, a natural Umwelt is transformed into a symbolic ecosystem. As Lacan phrases it, psychoanalysis is "creationist" (p. 667). It documents how a symbolic environment is *created* by technoscientific contrivances. Psychoanalysis itself likewise "operates in the symbolic" (p. 677).

Gadgets

During the academic year 1969–1970, while Jacques Lacan presented his famous *Seminar XVII*, Marxism was in the air, eagerly adopted by Maoist students and many of their teachers, arguing that the global proletariat should gain political control. Many of these students considered psychoanalysis a bourgeois practice, focusing on familial dynamics in affluent bourgeois families, using iconic Greek mythology and tragedy (i. e., Oedipus as gymnasium course material) as a frame of reference.

For Lacan, however, who himself came from a bourgeois catholic background, the traditional Marxist aim of establishing a proletarian dictatorship, and eventually a communist society, seemed a questionable objective. Maoists, he argued, fail to notice that something has changed: a transition or mutation *within the symbolic order as such*, which outdated the traditional Marxist view on technology (p. 207). What struck him as a distinctive feature of Maoism was the emphasis on handbook knowledge, provided by "manuals", exemplifying the manual knowledge of the exploited. Lacan saw Mao's red booklet as a prototypical political manual for producing revolutions. Yet, he argued, something completely new has emerged in our world, little things called "gadgets", in which technoscience now objectifies itself – entities that are entirely forged by science. And now the question is whether, in such a high-tech world, completely under the sway of information technologies, manual know-how can still carry sufficient weight to count as a subversive factor (p. 174). Manuals were written in order to operate machines, but for Lacan gadgets exemplify a completely new type of machine, involving brain work, information science, robotics, and cybernetics, rather than manual labour.

Gadgets, as an unprecedented technological phenomenon sui generis, not only pose a challenge to Marxism, but to Freudian psychoanalysis as well. Freud had famously claimed that, contrary to fear, which is object directed, "anxiety" is without an object. Lacan now disagrees with this and claims that anxiety is instilled precisely by these little interconnected objects, which he refers to objects *a*, tending towards invisibility (*Verborgenheit*), but pervasive and omnipresent (p. 172, p. 216), putting us under constant surveillance, and therefore generating anxiety. Big Brother is watching us. The imperative of this new symbolic order generated by gadgets is: "Work harder!" but at the same time, "Enjoy life to the full!" – "Never enough!"

Thus, in the global society of late capitalism, science is producing a new type of entity, functioning as objects of anxiety and desire, as objects *a*, namely electronic gadgets, pervading the lifeworld with their unnoticeable, vibrating, Hertzian waves, relying on the manipulation of symbols, creating a new, artificial environment, which Lacan refers to as the "alethosphere" – Lacan's version of what Teilhard de Chardin referred to as the "noosphere" (Zwart 2022b). Things like microphones connect us to the alethosphere, and even astronauts floating in space, Lacan argues, although they have left the geosphere and the atmosphere, are still connected with the alethosphere, with "Houston", via gadgets, representing the human voice, but in an inorganic version, detached from the body, as object *a* (p. 188). The world is increasingly populated by these gadgets (p. 185), these tiny objects *a*, which we encounter everywhere, in institutional buildings and in shopping malls, pervading the global metropolitan environment (Zwart 2020b).

Lacan refers to these gadgets as "lathouses", a jocular, mock Heideggerian portmanteau term, combining "ousia" (being) with "aletheia" (truth) and oblivion ("lethe") to indicate how these pervasive technological entities proliferate as entities (*ousia*) designed to disclose the human world by creating a technological clearing (*aletheia*) while we tend to forget and overlook their presence (*lethe*). First and foremost, however, they are objects of jouissance, allegedly designed to satisfy our desires, but proving unsatisfactory in no time. Consumers want them, they desperately need them, and they desire the latest versions of them, but in the end these electronic commodities exploit and consume their consumers, rather than the other way around, continuously registering and disseminating information, without us being aware of them (Millar 2018; 2021). They produce and circulate data on a massive scale, functioning as neo-liberal knowledge concerning what consumers want. Human desire and human self-determination are operationalised in terms of click-behaviour, captured by search algorithms. Via these gadgets, algorithms are running wild in the world (Possati 2020). Although we usually cannot directly see them, we intuit their presence, so the idea that we are surrounded by gadgets causes anxiety. While allegedly supporting pleasure and freedom, they de facto confront consumers with normalcy standards and societal expectation, informing them that they must work harder on themselves and keep a close watch on themselves, to optimise psychic and somatic functioning and to postpone the impacts of unhealthy lifestyles (e.g., competitiveness and stress) and ageing. Let me now further elucidate this view with the help of a case study.

A case study: *Homo Deus*

In *Homo Deus*, Yuval Harari (2015/2016) presents the history of *Homo sapiens* as a tantalising success story: amounting to a "conquest of the earth" by one singular species. This conquest has recently experienced a dramatic acceleration, moreover, due to the rapid production, accumulation, and distribution of knowledge through artificial intelligence and digitalisation. The capacity of humans to outcompete other species, Harari argues, is due to our intelligence, which entails an unprecedented capacity to adapt. Yet, Harari also notices a number of emerging challenges. First of all, our dramatically increased capability to gratify human needs on a global scale paradoxically gives rise to a no less dramatic increase and proliferation of needs and expectations and, in the end, to more unhappiness. Secondly, our anthropocentric focus on satisfying human needs has resulted in a dramatic destabilisation of the global ecological equilibrium. Finally, artificial intelligence is bound to replace and outcompete humans, so in the near future, large numbers of humans will become superfluous. We are not at the final stage of history, but rather represent a stage in world history that is about to be overcome.

The paradoxical experience that the rapid increase of wealth and productivity leads to more unhappiness, Harari argues, is the fault of our biochemical system (p. 43). We all want lucrative jobs and good-looking partners, but essentially, even Don Juans and business tycoons are nothing but lab rats seeking more gratification by pressing pedals (p. 44): never enough. The only way to solve this would be to biochemically reengineer *Homo sapiens* so that we can enjoy everlasting pleasure (p. 49), as high-tech lotus eaters.

As Harari points out, artificial intelligence is not a mere instrument allowing us to make human existence happier and more efficient. Rather, AI seems to be *using us* to advance itself. There is no reason to think that *sapiens* is the final station of evolution. Biochemical neural networks are already being replaced by intelligent software as we speak (p. 52). Humans themselves are becoming the prime target of this development, changing one feature after another, from hip implants to transexual surgery. All partial organs or subsystems will prove replaceable so that we not only reassign our gender but also become more bionic and machine like, until we are no longer human. *Homo sapiens* will disappear, and human history will come to an end. And if we prefer to stay the way we are, robots and other intelligent machines will soon replace us. Compared to intelligent machines, human beings have developed many features that now prove to be superfluous, such as consciousness. And even morality is nothing but a set of modifiable algorithms.

Homo Deus entails a philosophical anthropology: an answer to the question of what we are as human beings. According to Harari, artificial intelligence informs us that all organisms are basically algorithms (p. 96 and passim). An algorithm can be defined as a methodological set of steps to make calculations, resolve problems, and reach decisions, and organisms are a vast collection of methodological sets of algorithmic steps, developed, refined, and tested in the course of evolution. This applies to humans as well: humans are algorithms; their thoughts, sensations, and emotions are algorithms; and even their desires are "highly refined algorithms" (p. 102). We are not a holistic entity but an evolving assemblage of partial organs (121), functioning on the basis of algorithms. Humans are "craving beings" (p. 124), but even our sensations, wishes, and emotions are biochemical data-processing algorithms, electrochemical reactions in the brain (p. 124).

One of the anomalies in Harari's understanding of humans is the existence of subjective feeling or consciousness. This phenomenon seems superfluous and redundant. What could be the evolutionary benefit of consciousness, of a self-conscious ego? Why add subjective experience to an algorithmic biochemical machine? Harari admits that this is a lacuna in his understanding. Ninety-nine percent of our bodily activities take place without conscious feelings and mathematical symbols can describe virtually all the algorithms of the human brain. None of the computers or data-processing systems we have created need subjective experience to operate. No computer program needs consciousness, desires, or drives. Why then have these psychic phenomena evolved? Most scientific research fields have deleted consciousness from their vocabularies, seeing it as a by-product of brain processes: a kind of "mental pollution" produced by the firing of complex neural networks (p. 136), although paradoxically, we still build our whole edifice of politics and ethics upon something which rational and scientifically informed individuals can no longer believe in, namely subjective experience. Many scientists still claim to believe in things like autonomy, freedom, and God, although they hardly ever write about such concepts in their scientific papers. Why then are these concepts retained? This also applies to the basic concepts of psychoanalysis.

As Harari phrases it, “our jargon is still replete with Freudian psychology, but our computers don’t crave when they have a bug” (p. 136).

Homo Deus also entails a view of human history. In the past, Harari argues, *Homo sapiens* “used language to create new realities” such as nations, demons, and gods (p. 175), but there is no longer any need for such fictitious entities, he contends. They may have played a role in the past but are now becoming obstacles to progress. Human history (e. g., historical phenomena such as the Crusades or Communism) can now be explained in strictly biochemical and algorithmic terms. Indeed, the concept of the algorithm allows Harari to rewrite the annals of world history altogether. An important turning point came about, he argues, when in ancient societies, language gave rise to writing: a technology which enabled humans to organise entire societies in an algorithmic fashion (p. 187). Ancient Egypt and ancient China were created in this manner. Thus, local verbal rituals (analphabetic algorithms) were replaced by the algorithms produced and applied by a literate elite. Via humans, the algorithmic principle has reshaped the world, starting with ancient societies such as ancient Egypt and China and their literate bureaucracies, using script to design methodical procedures to make calculations, resolve problems, and reach decisions. While Egyptian Pharaohs and Chinese Emperors were basically “imaginary entities” (191), their bureaucrats, and the algorithms they employed, really mattered. Scribes existed, but kings and nations existed only in the imagination.

Inevitably, these algorithms collided with reality, and they still do. This means that either reality will have to be rectified and refurbished, or our algorithms must be refined, or both. In the present, due to computers and bioengineering, the difference between algorithms and reality begins to blur to the extent that reality itself becomes increasingly algorithmic, i. e., refurbished by algorithms. The newly designed artificial algorithmic contrivances of technoscience build on and interact with the packages of algorithms that evolved through biological evolution. Until recently, most societal systems emerging in history aimed to establish and maintain an equilibrium with the help of tested algorithms (“traditions”), but Western modernism is deviant. Instead of trying to maintain an equilibrium, modernism intrinsically strives for acceleration and exponential growth, disrupting all equilibriums it encounters along the way. Throughout history, monarchs relied on tested algorithms while wasting their surplus capitals on flamboyant carnivals, sumptuous palaces, and unnecessary wars, but modern capitalism relentlessly invests in growth, so we now live in a world of chaos, disturbance, and uncertainty. And the symbolisation of the real has resulted in a global situation where the most important resource is something purely symbolic, namely digitised algorithmic knowledge.

And now, even humans themselves are pushed into the defensive. On the job market, AI outperforms humans in most if not all cognitive tasks. In the near future, this will result in a massive class of economically "useless" people. To some extent, communism already embodied the same logic, allegedly still striving for equilibrium, but projecting it onto an imaginary future. Marx and Lenin studied how steam engines functioned, how coal mines operated, how railroads shaped the economy, and how electricity influenced politics. No communism without electricity, Lenin once argued. Contemporary capitalism diverts from its predecessors, Harari contends, because it relies on computer algorithms. If Marx came back to life today, he would probably urge his remaining disciples to devote less time to studying *Das Kapital* and more time to studying the Internet and the human genome (p. 319). Although we can learn something from history, the focus should be on algorithms here as well. This would allow historians to discern, for instance, that the Catholic Church was basically an algorithmic organisation, establishing medieval Europe's most sophisticated administrative system, using inventions such as archives, catalogues, timetables, and other techniques of data processing. According to Harari, the Vatican was the closest thing twelfth-century Europe had to Silicon Valley. The Church created Europe's first economic corporations – the monasteries – which for thousand years spearheaded the European economy and introduced advanced agricultural and administrative methods. Monasteries were the first institutions to use clocks, and for centuries they and their cathedral schools were the most important learning centres of Europe, helping to found universities. Yet, if we want to understand the present, it is more important to understand genetic engineering and artificial intelligence. To come to terms with the present, we should read scientific articles or, even better, conduct lab experiments ourselves, instead of debating ancient texts.

In the present, humans are losing control. Brain scanners can now predict our decisions and our desires, revealing that the latter are nothing but algorithmic patterns of firing neurons (p. 333). The difference between humans and computers is that humans have something extra, namely a narrating ego, producing plots full of imaginary entities, lies, and lacunas that are rewritten again and again. Computers and robots do not need such a self-deceiving device. They have decoupled intelligence from consciousness (p. 361). They are intelligence without an ego and will soon outperform humans in all cognitive tasks. Non-conscious intelligence is about to bypass consciousness, giving rise to the superintelligence of global networks. Computer algorithms will soon acquire a monopoly over all key sectors of society including traffic control, the stock trade, travel agencies, health care, and pharmacy, optimising algorithms continuously with the help of statistics garnered from trillions of events. *Homo sapiens* is an assemblage of algorithms shaped by evolution, so most human qualities are becoming redundant when it comes to performing modern jobs. Let this suffice as a summary of Harari's views. In the next section, I will assess the logic of his approach from a Lacanian perspective.

Algorithms and desire

At first glance we may notice some remarkable concordances between Harari's algorithmic logic and Lacanian psychoanalysis. From a Lacanian perspective, technoscience indeed entails a symbolisation of the real, an endeavour which only works because the real is already intrinsically algorithmic, consisting of constituents (elementary particles, nucleotides, genes, etc.) that can be present or absent and can be referred to with alphabets of symbols (στοιχεῖα) which function as basic constituents of algorithmic programs. Basically, two stratagems for coming to terms with the Real are open to humans, as we have seen, the Imaginary and the Symbolic. In the first case, narrative egos produce and invoke fictitious entities (gods, nations, myths, etc.) but technoscience basically opts for the second stratagem: symbolisation. The replacement of imaginary ideas by symbolic (algorithmic) interpretations is a basic iconoclastic tendency at work in the history of human civilisation, while digitisation can be considered the highest stage of these processes of symbolisation. Here, symbolisation and its programs are finally reduced to their bare essentials: presence or absence, 1 or 0. Symbolisation, notably in this radical form (through algorithmicising and digitising), dismantles the imaginary, the realm of mythical and fictitious entities. Icons are replaced by algorithms.

At the same time, we should notice, from a Lacanian perspective, that notwithstanding this iconoclastic fervour at work in history, the symbolic dimension will never completely succeed in eliminating the imaginary altogether. Rather, Lacanian psychoanalysis points out that every advance in symbolisation will generate new icons (e.g., powerful images produced by technoscience such as the double helix or the Big Bang). Every algorithmic system always operates in both registers. This also applies to the examples mentioned by Harari such as the literate bureaucracies of ancient China and the medieval Catholic Church. The administrative rituals and practices of the medieval Church produced and relied on icons as well as on algorithms. Symbolisation is compensated by a return of the repressed (a negation of the negation). This also applies to Yuval Harari himself, as will be argued in more detail below. He himself is also a narrating ego whose effort to explain world history in terms of algorithms results in a narrative which contains some decidedly imaginary elements, but we will come to that.

Ancient Greek philosophy introduced the science of dialectic (the logic of syllogisms) in a world replete with imaginary entities. Greek and Roman statues, for instance, embodied an iconic view of human beings, notably of human *bodies*, erected at crossroads and other strategic places as a form of moral propaganda: this is our telos, this is what we should aim to become (1955–1956/1981). Due to technoscience, such iconic images are now replaced by algorithms. We are turning ourselves into quantified selves. This indicates that, to some extent, we humans *really are* an aggregate of algorithms. For Harari, however, this insight is connected with the implicit conviction that human-made algorithms tend to function rationally and smoothly. What is decidedly absent in Harari's worldview is the daily experience that the algorithms we encounter or produce are chronically dysfunctional. The present world consists of a plethora of algorithms, but discontent in digital civilisation stems from the basic experience that most if not all of these algorithms fail to function in the end. They are disruptive rather than smooth and functional.

Therefore, Lacanian psychoanalysis suggests at least one correction to Harari's viewpoint: desire is not an algorithm, and it is not even a "highly refined algorithm" (p. 102). Desire points to a disruption in the algorithmic system *as such*. Psychoanalytic concepts such as discontent and desire point to the experience that most if not all algorithms developed by civilisation prove dysfunctional in the end. While purportedly designed to solve our problems, they inevitably create increasing amounts of chaos, frustration, entropy, and disruption as well. Instead of "unhappiness", a more optimal term for this type of dysfunctionality – this lack of concordance between psyche and world, between internal and external algorithms, between what algorithms purport to deliver and how they actually operate – is discontent (*Unbehagen*), a concept which refers to the discomfort and unease experienced in civilisation: a system that claims to satisfy all our biochemical needs, but without removing the susceptibility of all human endeavours to failure. Desire is never a mere bug; it is not something that can be eliminated through optimisation. Rather, it points to a basic disruptive disparity. Although civilisations may to a large extent satisfy our biological needs, they also introduce expectations and demands which we, as deficient craving beings, will never be able to live up to (e.g., be more competitive, be more productive, more innovative, enjoy life to the full, never enough, etc.).

Precisely for this reason, Harari argues, humans are bound to become replaced. If humans are replaced by algorithmic machines, desire and discontent will allegedly become something of the past. Harari's confidence in the rationality of algorithms and algorithmic organisations is not supported by evidence, however, to put it mildly. Rather, his imaginary idea of a perfect algorithmic device functions as a deceptive fantasy. All algorithms prove dysfunctional in the end. I fully agree with Harari that philosophers in general and Lacanian philosophers in particular should be encouraged to study crucial technoscientific developments such as genomics and the Internet, but they should also be encouraged to study the literate bureaucracies of ancient China or the administrative algorithms of the medieval

Catholic Church. If we study the Internet, we will notice how this infrastructure actually fostered global polarisation. And if we thoroughly study the history of algorithmic organisations such as empires and churches, we will notice that they actually find themselves in a permanent state of crisis. This also applies to the university as an algorithmic institute whose fascinating history reflects a permanent state of crises (cf. the countless publications with the signifier "crisis" in the title that appeared since modern universities were installed). Both the Chinese empire and the Catholic Church aimed to establish an everlasting equilibrium, by offering a scaffold for human desire, but this only worked to the extent that they allowed sufficient space for strategies of sublimation, e. g., for the poetic and the iconic. Algorithmic organisations such as the Chinese empire and the Catholic Church created worlds of architecture, poetry, and learned treatises because they managed to reconcile the symbolic and the imaginary to some extent ("sublimation").

Freud's label for the logic of the algorithm is the pleasure principle. The pleasure principle could be seen as an algorithmic program to optimise satisfaction. Yet, the key discovery of psychoanalysis was that human desire is unquenchable because it takes us beyond the pleasure principle (beyond the logic of optimisation or gratification of biochemical needs), pointing to a disruptive and entropic dimension in the functioning of algorithms (thematised by Freud as the death drive). The idea of an optimised world consisting of algorithmic machines replacing humans is fictitious because it obfuscates the disruptive and entropic impact of all algorithmicising practices. This applies to the AI algorithms that govern the current neo-liberal stock market as much as it applied to the literate bureaucracies of ancient China. Their entropic impact gives rise to discontent and instances of profound crisis. They are prone to anomalies and inner contradictions which are not reducible to repairable bugs. Their logic rather proves self-destructive in the end.

Although the concept of the algorithm as such is an interesting intellectual device, it deserves a more dialectical handling. Algorithmic programs evolve, not via optimisation, but in a more dramatic fashion, by the recurring experience of paralysing inner contradictions, which give rise to anxiety and discontent and can only be superseded by recognising that the initial view was driven by imaginary and ideological components, such as the fictitious idea of a smoothly functioning future world of algorithms. When this idea is pushed to its extreme, as happens in contemporary neo-liberalism (celebrated in Harari's bestseller), its inner contradictions are bound to be revealed in a dramatic fashion, in the form of global disruption. Exposed to the real, the validity of this logic is negated, and this negation can only be superseded (negated) by using this experience to develop a more comprehensive view which acknowledges that, left to its own devices, the logic of algorithms easily becomes disruptive. In short, Harari, the author himself, is a narrating ego producing a plot that is full of imaginary entities and lacunas.

Hegel himself already agreed that “everything is a syllogism” (Zwart 2022a, p. 41), but at the same time he acknowledged that a real syllogism evolves in a dramatic fashion, through negation, frustration, and refutation. Algorithms are tested, refuted, and challenged rather than applied and optimised. Algorithms are both beneficial and disruptive. They foster the hope of optimisation but only work if their entopic and disruptive tendencies are duly recognised and addressed. If not, they inevitably result in symptoms of alienation and dehumanisation. To provide a convincing assessment of the past, present, and future of global civilisation, Harari’s ideological conceptualisation of the algorithm requires a thorough Lacanian (i. e., psychoanalytical-dialectical) rereading.

References

Aydin, C. 2021. *Extimate technology: Self-formation in a technological world.* New York: Routledge.

Ellenberger, H. 1970. *The discovery of the unconscious.* New York: Basic Books

Fink

Freud, S. 1938/1941a. *Abriss der Psychoanalyse.* Gesammelte Werke XVII, 59–62. London: Imago.

Freud, S. 1938/1941b. *Die Ichspaltung im Abwehrvorgang.* Gesammelte Werke XVII, 63–140. London: Imago.

Freud S. 1913/1940. *Totem und Tabu.* Gesammelte Werke IX. London: Imago.

Freud S 1917/1947. *Eine Schwierigkeit der Psychoanalyse.* Gesammelte Werke XII, 3-12. London: Imago.

Freud S. 1920/1940. *Jenseits des Lustprinzips.* Gesammelte Werke XIII, 1–70. London: Imago.

Freud, S. (1930/1948) *Das Unbehagen in der Kultur.* Gesammelte Werke XIV, 419–506. London: Imago.

Freud, S. (1950) *Aus den Anfängen der Psychoanalyse. Briefe an Wilhelm Fliess, Abhandlungen und Notizen aus den Jahren 18871902.* London: Imago

Freud. S. 1910/1943. *Über den Gegensinn der Urworte.* Gesammelte Werke VIII. London: Imago.

Harari, Y. 2015/2016. Homo Deus: a brief history of tomorrow. London: Vintage / Penguin.

Lacan, J. 1953–1954/1975. *Le Séminaire I : Les Écrits Techniques de Freud.* Paris : Éditions du Seuil, 1975.

Lacan, *J.* (1955–1956/1981) *Le Séminaire III : Les psychoses.* Paris : Éditions du Seuil.

Lacan, J. 1959–1960/1986. *Le séminaire VII : L’éthique de la psychanalyse.* Paris : Éditions du Seuil.

Lacan, J. 1960/2005. “Discours aux Catholiques.” In: Lacan, J. *Le triomphe de la religion. Précédé de discours aux catholiques.* Paris: Éditions du Seuil, pp. 9–65.

Lacan, J. 1961–1962. *Le Séminaire IX : L’identification* (unpublished). http://www.valas.fr/.

Lacan, J. 1963/2005. “Introduction aux noms-du-Père”. In: *Des Noms-du-Père.* Paris: Éditions du Seuil, pp. 65–104.

Lacan, J. 1964/1973. *Le Séminaire XI : Les quatre concepts fondamentaux de la psychanalyse.* Paris : Éditions du Seuil.

Lacan, J. 1966. Écrits. Paris : Éditions du Seuil.

Lacan, J. 1974/2005. “Le triomphe de la religion.” Lacan, J. (2005) *Le triomphe de la religion. Précédé de discours aux catholiques.* Paris: Éditions du Seuil, pp. 67–102.

Millar, I. 2018. Black Mirror: from Lacan’s lathouse to Miller’s speaking body. 36 (2).

Millar, I. 2021. The Psychoanalysis of Artificial Intelligence. London: Palgrave.

Possati, L.M. (2020) Algorithmic unconscious: why psychoanalysis helps in understanding AI. *Palgrave Communications.* DOI: 10.1057/s41599–020–0445–0.

Sulloway, F.J. (1979/1992) *Freud, biologist of the mind: Beyond the psychoanalytic legend.* Cambridge / London: Harvard University Press.

Žižek S. 2016/2019. *Disparities.* London: Bloomsbury.

Zwart, H. 1995. "Ik ben een machine... Over het Cartesiaanse gehalte van de Freudiaanse Zelf-conceptie". In: H. Hermans (red.) De echo van het ego: over het meerstemmige zelf. Annalen van het Thijmgenootschap, 83 (2). Baarn: Ambo, pp. 49–80.

Zwart H. (2013) The genome as the biological unconscious – and the unconscious as the psychic 'genome'. A psychoanalytical rereading of molecular genetics. Cosmos and History: the Journal of Natural and Social Philosophy 9 (2): 198–222.

Zwart, H. 2016. The obliteration of life: depersonalisation and disembodiment in the terabyte age. *New Genetics and Society* 35 (1) 69–89. DOI: 10.1080/14636778.2016.1143770

Zwart, H. 2019. *Psychoanalysis of technoscience: symbolisation and imagination.* Series: Philosophy and Psychology in Dialogue. Berlin/Münster/Zürich: LIT Verlag. ISBN 978-3-643-91050-9. Series: Philosophy and Psychology in Dialogue. Volume 1.

Zwart, H. 2022a. *Continental philosophy of technoscience.* Series: Philosophy of Engineering and Technology. Dordrecht: Springer

Zwart, H. 2022b. The Symbolic Order and the Noosphere: Pierre Teilhard de Chardin and Jacques Lacan on technoscience and the future of the plane. International journal of Philosophy and Theology. 10.1080/21692327.2022.2093775

Kerrin A. Jacobs

(Nothing) Human is Alien – AI Companionship and Loneliness

Abstract: AI companionship promises a new way of coping with loneliness experiences in highly digitalised societies. In a first step some basic criteria that characterise the relationship with a companion AI (social x-bots) as distinct from human relatedness are sketched. While AI companionship is often praised for the potential to cope with loneliness its crucial flaw is its lacking of an intersubjective dimension, which is essential for the human condition. My hypothesis is that AI companionship cannot solve the problem of loneliness, which is elaborated in a second step. Against the background of theories of social relatedness and a description of loneliness as a form of alienation, digital loneliness is respectively emphasised as a new form of cultural discontent. It paradigmatically reveals the "echo chamber scenario" of the human-x-bot companionship and therefore cannot be cured within this constellation. The x-bot companion rather constantly reminds the lonesome of that what cannot be substituted by or compensated for by AI: namely, social recognition. The conclusion is that the strategies of humanising AI come with the cost of dehumanising companionship. Digital loneliness symptomatically reveals both: that what never is alien and simultaneously will always stay alien to humans.

Keywords: AI companion, loneliness, alienation

1 Introduction: (Nothing) human is alien

Homo sum, humani nihil a me alienum puto.[1] This saying seems to be quite apt to describe loneliness, which, especially in recent times, has even been ascribed to the status of a new widespread disease (Spitzer, 2018). We can acknowledge that it can affect us all, that it is a kind of abandonment and alienation that painfully strikes at our primordial human core: our nature as relational beings. Loneliness is a coming of awareness of the old hidden fear of being abandoned, rejected, and feeling misunderstood or not recognised and is never alien to us as a possible way of suffering from our existence. It reappears in its most uncanny form in the com-

1 "I am a human being, nothing human is alien to me" is a winged word from the comedy *Heauton Timorumenos* ("The Self-Tormentor") by the poet Terence (verse 77). Cf. Lefèvre (1986, p.39–49).

https://doi.org/10.1515/9783111007564-004

panion relation to an x-bot[2]. The x-bot is what appears familiar but is also creepy as it reminds humans of that which is essentially missing or is always feared to be lost: the feeling of being related to others and recognised. Loneliness is "the other" threatening dimension of social relations we can normally repress. That now of all things x-bot companion in its whole (un-)familiarity (re)appears as a solution for coping with loneliness does not miss a certain irony but is also appealing with regard to the culture-analytic question of what it means to be a human being, or rather, to be a lonely human.

2 Digital Loneliness – The recall of discontent

What is lost in loneliness is nearness to others.
Hans-Georg Gadamer (1988)

It should be mentioned right from the beginning that I exclude more complex virtual strategies for loneliness prevention where real persons help lonely people with virtual scenarios to get in touch with other people but focus rather on the one-to-one relation of AI-human companionship.[3] The relationship between humans and companion AI (social x-bots) is characterised by many possibilities. Everything that is possible demands limits, since this is the precondition for realisation. Such limits are set with the functional design of the respective x-bot, while the experiential possibilities that companion AI, for instance, those that appear in a human-like shape, such as holograms like *Hatsune Miku*[4] or companion (love) dolls like *Harmony*,[5] are realised within the limits of the human mind, thus also are limited by specific (moral, religious, socio-political, ethical) norms

2 In the following I use the expression "x-bot" to address companion AI or "sociable" robots (as they are called, e. g., by Breazeal and Scassellati 1999), well aware that there are significant differences between robots, such as Winky, Aibo, or Sofia and conversational agents, chatbots, such as ELIZA, Alexa, Xiaoice, Replika, Tess, Woebot and Wysa, etc.

3 Promising scenarios for effectively coping with loneliness are given with the opportunity to meet *fellow-beings* in a virtual reality space that, e. g., allows elderly people to contact their friends and family. See: Barbosa Neves, Waycott and Maddox (2021); Johnston (2022).

4 Miku Hatsune is a virtual character designed by Japanese mangaka and illustrator Kei Garo on behalf of Crypton Future Media, which the company Gatebox uses as a character for their companion chatbot.

5 Harmony is a RealDoll companion robot that allows customers to create their doll in looks as well as in personality according to 10 "persona points" such as how sexual, moody, or intense the doll should be. https://www.althumans.com/companion-robots/real-doll.html (accessed 24.11.2022).

and values. The relationship to x-bots can be called *real* to the extent that it certainly contains a sort of horizon (cf. Husserliana XXXIX; Jorba, 2020; Geniusas, 2012)[6], a perspectivity that is accompanied by a sense of experiential possibilities in interaction with the x-bot. The condition of loneliness itself can be described as a significant change of perspectivity – an existential change of horizon. If we problematise loneliness management with AI, it is predominantly about a change of perspectivity on how to be alone in the digital era, and what loneliness actually is today. The former gets addressed with the idea of loneliness, a significant change or even loss of meaningful relatedness to the world, others, or one's self. The latter is discussed by pointing out that what is (to date) not real, although it is psychic experienced reality – namely *re*cognition by a machine – traps the lonely in an AI echo chamber, in which they eventually might even lose sight of the potentially problematic nature of this monologic existence. One might oppose that using a technical device like an x-bot for working through one's loneliness all alone might be quite the opposite of being trapped, but rather an authentic expression of (digital) autonomy. But is this really a good way of coping with it? Does it demonstrate being "in control" of one's loneliness, or even enable agents to actually leave it? This can be doubted, at least, when individuals stay with a psychic reality that is cut off from the material reality of intersubjective practice, as the decisive realm in which agent autonomy becomes visible and effective as the culturally hard-won adaption to the material reality of the social.

2.1 AI-companionship – A canny cure for loneliness?

We are all robots when uncritically involved with our technologies.
Marshall McLuhan (2017)

The trajectory of robotics over the last 50 years, starting with prototypes like *Shaky*[7] and ending with high-end AI of human-like appearance like *Sofia*[8] is, in-

---

6 In Husserl's phenomenology the horizon is a general structure of experience (Cf. Husserl, 2008).

7 Shaky was the first general-purpose mobile robot able to analyse commands and break them down by itself. It was developed in 1966–1972 at SRI International and the project was funded by the Defense Advanced Research Projects Agency (DARPA).

8 Sofia is a social robot developed by the Hong Kong–based company Hanson Robotics and is exceptional as in 2017 it was the first robot to receive citizenship and was the first non-human to be given a United Nations title (UND Programme's first Innovation Champion).

deed, fascinating. AI companions, like the cat-like robot *Winky*[9] or chatbots, can be easily archived by the ordinary customer. The particular relationship horizon it offers follows the objectives and purpose-bound needs of our nature and can be relatively easily integrated into the particular design of, e.g., our communication settings. Apparently, feeling less lonely is a desire to which AI companions are perceived to offer a solution with canny design, which allows for the development of a relationship between humans and x-bots (Clark et al., 1999). Studies, particularly on chatbot use (cf. Skjuve et al., 2021), show that certain AI is even perceived as a family member or friend (Gao et al., 2018; Purington et al., 2017)[10]. Moreover, the more humane AI appears, the more likely we are *perceiving* it as an object of patience, and because an immoral treatment of x-bots may have harmful consequences to human interactants, as, for instance, Cindy Friedman (Friedman, 2020, p.10 ff) suggests.[11] In addition to its being an object of patience, one certainly cannot doubt that the relationship of persons to their personalised robots is a libidinous one. The x-bot is a *beloved object* (cf. Habermas, 2020), a *constantly available object*, and an *object of projection* and *projective identification* (Szollosy, 2017, p.435)[12]. Interactions with it lack inter*subjectivity* which rules it out as "equal" to human interactants, albeit we relate to them. This has inspired Sherry Turkle (Turkle, 2004) to claim the need for a new objects relation theory according to which relational artefacts can be ascribed the role of self objects. Originally stemming from Heinz Kohut's idea that "some people may shore up their fragile sense of self by turning another person into a 'self-object'" (Turkle, 2004)[13] x-bots "clearly present themselves as candidates for such as role […] If they can give the appear-

9 Winky is a play-bot of the French start-up Mainbot and has microphones, sensors, a speaker, LEDs, a rotating head, rotating ears, a motion and distance detector, and a gyroscope that allow it to interact with children and the environment.

10 After some disappointment with women, a Japanese man called Akihiko Kondo even married the virtual figure Hatsune Miku and till today seems to be puzzled by the fact that his partnership with a hologram is not accepted by others as a "real" partnership. See: https://www.otaquest.com/hatsune-miku-gatebox-marriage/ (accessed 25.11.2022).

11 This is a central topic in the ethics of human-robot interaction. Cindy Friedman claims that the perception of robot consciousness (she stays neutral on that) and moral patience together with the possibility of treating a robot immorally has the consequence that humans harm themselves with such transgressions towards social robots, as the human is *always both* moral agent and moral patient of their moral actions towards a robot, while social robots are only *perceived* moral patients (Friedman, 2020).

12 For Szollosy thematised the projection in terms of the paradigm of the "imagined monster robot", i.e., the robot as a container for all our anxieties and fears, and of ourselves as rational agents, in his analysis of the figure of the robot as a monstrous machine in popular media (Szollosy, 2017).

13 Here, Turkle refers to Kohut (1978).

ance of aliveness and yet not disappoint, they may even have a comparative advantage over people [...]" (Turkle, 2004, p.20). Moreover, if humans were equally seen as available like x-bots, this would be the opposite of recognition – namely reification – thus, a severe form of dehumanisation. And even if AI would meet the basic pre-requirement for inter*subjective* practice, the sociability of intersubjective practice between humans is (to date) still fundamentally distinct from and exclusive in sense-making from that with a social x-bot.[14] The way meaning is enacted (renegotiated, acquired, confirmed etc.) in the intersubjective modes of relatedness of self-reflexive beings that are enabled for *re*cognition is out of reach even for high advanced companion-x-bots. Recognition of the other not in *technical* terms (of pattern recognition), but in *ethical* terms of social inclusion cannot be provided by an artificial agent in any comparable "social" way than moral agents can provide it. Unsurprisingly, exactly this is what the companion-x-bot design (un)cannily "simulates": The impressive illusion of affective attunement, mutual respect, and understanding of the other. This is evidenced by studies that describe that humans have developed – what they would call – trustful and affective relations to chatbots (cf. Skjuve et al., 2021). The main facilitating factors are that the AI is perceived as non-judgmental and accepting and displays "understanding" to the user. However satisfactory this continuous support might be experienced, if the psychic reality of a feigned intersubjective engagement with a (people pleasing) companion AI overturns the critical faculty of *reality testing* (Freud, 1911)[15] – e.g., forgetting to remember oneself of the fact that one factually is still lonely

---

14 The talk of AI "consciousness" seems to date still only a metaphor, albeit the idea of an analogy between the human consciousness and the algorithmic activity of AI is thought provoking, especially for psychoanalysis. Luca Possati has exemplarily pointed this out with the central hypothesis that the method and concepts of psychoanalysis can be applied in order to better understand human-AI interaction, thereby stressing in particular on the notion of the unconscious, which he re-interpretated in technological terms, and elaborates on how we can add a new register to the symbolic and imaginary identification with AI. See: Possati, L. (2021). I also would like to recall Lacan's project of discourses and the unconscious, in which he has taken the comparison between man and machine very far. What has to be noticed is that this simply comes to an end where one may assume that the information-theoretical correlate of the repetition compulsion is a suitable analogy but corresponds to a "beyond" of the pleasure principle, ergo referring to the beyond of meaning, but it is meaning what we are aiming to gain in psychoanalysis (cf. Langlitz, 2005, p.193 ff).

15 Classical psychoanalysis refers to the reality principle as part of the ego. It forms a pair with the pleasure principle, which is modified by the reality principle inasmuch as the aspirations of the pleasure principle, which originate from the id, are adjusted according to the requirements of the environment. Reality testing refers to the function (and psychotherapeutic technique) by which the internal world of thoughts, feelings, and desires becomes distinguished from the external (objective) world. Cf. Freud (1920).

– one may ironically get trapped in a companionship that is rather a symptomatic expression of loneliness than an adequate cure for it. From a developmental perspective it can be additionally mentioned that the great integrative achievements of ego functioning are, besides a stable value orientation, a mature defence, particularly to cope with criticism, incalculable frustrations, refraining from the desire for immediate gratification, etc. Inasmuch as the x-bot companion "helps out" with uncritical blind affirmation and support and constant availability to serve our needs, this illusion of an intersubjective exchange calls for a reality double-check: The transformative possibilities of and in loneliness depend on a realistic assessment of one's situation, where the questions what loneliness means to oneself, how it signifies one's life, and what particular role a companion plays in it can become adequality addressed. I believe this is not possible with "relational artifacts" (cf. Turkle et al., 2006) and recalls the autonomous agent for actively seeking out for finding satisfying relations to other fellow-beings. Thereby one might face what makes one so afraid: the other, who is not the x-bot, but the other human beings to which connectivity, a shared horizon, has gotten lost. Nota bene: it might be the case that people find exclusively in x-bots something that they cannot share or even do not like to address in relations with their fellow-beings, and I believe this is also true when people relate to AI companions to cope with their loneliness (e.g., they do not have to explain themselves to an x-bot). Moreover, once AI becomes established and respected as a surrogate (e.g., by legal definition) this constitutes a relationship in its own right and may even be experienced as qualitatively superior to a relation with a human, e.g., with respect to fulfilling functions for which the respective AI has been selected. So, I do not rule out that it can be precisely the specific artificial nature, the "as-if" of intersubjectivity, which attracts humans to AI companions. In fact, the x-bot seems to help in the suppression of *feelings* of loneliness ("entertaining them away"), and particular modes of interaction (e.g., becoming very chatty and emotional with a social x-bot) can be considered a particular expression of *undoing*[16] (e.g., for

16 Freud first described the practice of undoing (German: *Ungeschehenmachen*) in his 1909 "Notes upon a Case of Obsessional Neurosis" (Freud, 1909). Undoing is a defence mechanism in which unhealthy, destructive, or otherwise threatening thoughts or actions are tried to be cancelled by engaging in quite the contrary behaviour. In 1926 Freud described the ego defence "[as] good enough grounds for re-introducing the old concept of defence, which can cover all these processes that have the same purpose –namely the protection of the ego against instinctual demands" (Freud, 1926, p. 324). Insofar as with Laplanche and Pontalis it can be stated that "[u]ndoing in the pathological sense is directed at the act's very reality, and the aim is to suppress it absolutely, as though time were reversed" (Laplanche and Pontalis, 1988, p. 478). Certain behaviour could be systematised under the auspices of suppression of loneliness.

not becoming actively engaged with other humans). The former, however, rules out not that someone *is* actually not lonely, and the latter repeats that idea: AI companionship is a paradigmatic expression of digital loneliness. While the canny design of x-bots helps to "deal" with the surface phenomena of loneliness, such emotional episodes as sadness, feelings of rejection, and boredom, and while they can make us forget our being alone, loneliness, however, cannot be overcome in and with AI companionship. It rather uncannily reminds us of the threats of a forgetfulness of recognition. This is now further explained:

2.2 Loneliness uncanny – The need for recognition

> (...) the more afraid you become of engaging with the world; and the less you engage with the world, the more perfidiously happy-faced the rest of humanity seems for continuing to engage with it.
>
> Jonathan Franzen (*How to Be Alone*, 2002)

Every strategy of humanising AI for coping with human "problems" such as loneliness has to reflect the difference between the vital and the mere active, between intersubjective sense-making and mere interaction, between the binary code that primes the algorithm of the machine and the complexity of evaluative modes of the human mind, between the capacity for recognition and that of cognition, and finally between the fact that people emotionally can suffer from and in relations with, and x-bots simply do not (although certain social artificial agents "display" some "emotion"). This might incline one to see x-bots not as sufficient *substitutes* for human companions and this is now further elaborated by highlighting basic qualities that unite all vital living beings: (1) we are relational beings[17], (2) we have some sort of striving and thereby transform the world into a place of salience, and (3) we can suffer from our particular modes of self-and-world-relatedness, sometimes in such a way that we become ill.

Loneliness always takes place in social embedding relationships, i. e., there is nothing such as loneliness without the sphere of the social: We can be (physically) alone without feeling lonely, and we can be around other people or AI companions and still feel lonely, so apparently it is the perceived quality of our relationships that determine whether we feel lonely. Loneliness is consequently a form of expe-

17 Søren Kierkegaard stated in 1844: "The self is a relation which relates itself to its own self, or it is that in the relation that the relation relates itself to its own self; the self is not the relation but that the relation relates itself to its own self". Der Begriff Angst (1844) /Die Krankheit zum Tode (1849) quoted here: Kirkegaard (2012), p.13.

rienced significant change or even loss of meaningful relatedness and this experience of alienation (from others, but also from one's self) encompasses the whole spectrum of thinking, feeling, willing, and acting of a lonesome person, and also has a bodily dimension. It is characteristic of autonomous agents that they actively constitute themselves in relation to their environment and therein sustain their integrity and identity. Humans, as relational beings, must mediate themselves with others and with their environments as a whole. Belonging to and being simultaneously also distinct from the social world are constitutive of the personal identity of a human being and are also an expression of the relationships in which an individual stands. These relational processes, coined "sense-making" (Thompson, 2007), capture all kinds of processes according to which an environment or being related to others presents itself as somehow salient or meaningful. Autopoiesis as "the all or nothing-norm of self-continuance" (Thompson and Stapleton, 2009, p.25) therefore is only a necessary, but not a sufficient, condition for sense-making, and adaptive autonomy is the central property of cognitive agents (Di Paolo, Rhode, and De Jaegher 2010) as "even the most simply organized systems regulate their interactions with the world in such a way that they transform the world into a place of salience, meaning, and value – into an environment (*Umwelt*)" (Thompson and Stapleton 2009, 25). Experiencing ourselves, others, and the world is an *inherently evaluative enterprise* bound to the procedural dynamics and (pre-)intentional character of consciousness. It is especially with respect to the notion of *participatory sense-making* that one literally can make sense of an agent's adaptive autonomy as deeply shaped by some sort of meaningful interaction with other agents: The other is perceived as a partner of mutual exchange in the *intersubjective* encounters of daily life (Fuchs and De Jaegher, 2009). The notion of participatory sense-making highlights the unique sensation of being (or rather: feeling of being) connected with each other. Interaction is a mode of social encounter that involves at least two agents who influence mutually each other through verbal and/or non-verbal behaviour in some time-dependent manner (De Jaegher, Di Paolo, and Gallagher, 2010; Di Paolo and De Jaegher, 2012; Ulrich et al., 2013; Schilbach et al., 2013) while full-blown *interactivity* is accompanied by a sense of "we-ness" (Gallotti and Frith, 2013) as an emergent property of interaction that provides the participants with or enables them to further acquire knowledge about each other and the proceedings of a certain situation. Crucial to the practice of sense-making with others is that we therein transcend a mere factual knowledge of the world and others in terms of a vital, immediate experience that plays a constitutive (albeit not exclusively explanatory) role in understanding other minds (De Jaegher, Di Paolo, and Gallagher, 2010; Di Paolo and De Jaegher, 2012).

Since being is "being-with-others" (cf. Heidegger, 1962)[18] loneliness always takes place in social embedding relationships. Consequently, theories that have emphasised loneliness as the "synthetic apriori of human consciousness" (see: Mijuskovic, 2012; 2015), according to which "the meaning of man is structured by a constant, futile struggle against his *isolated* state of conscious existence" (Mijuskovic, 2012, p.liii; italics KJ), must explain how they can make sense of the fact that "our fellow human beings are the most important part of environment" (James, 1884, p.195). Granted that another person's unique experience never can be entered, to conclude all experiences are structurally derivative, experiential sub-patterns of a more robust experiential pattern of loneliness, or mere surface-phenomena to an underlying "core essence of loneliness" becomes empirically challenged (cf. Jacobs, 2013b) with a developmental perspective on the brain: Basic modes of relatedness such as imitation, joint attention, and affective contagion are all processes of participatory sense-making, and crucial for the early development of the brain as a social organ (e.g. Fuchs, 2011; Papoušek and Papoušek, 1993; Colombetti, 2014). Only in these courses of embodied and meaningful interactions with the other can the neural systems, which are responsible for social cognition, mature. Accordingly, "[t]he resulting specialized neural networks should best be regarded as components of overarching interaction cycles: once formed in the course of these interactions, they serve as open loops for future situations presenting similar requirements to the individual" (Fuchs, 2011, p.209). This means: "participation" is prior, or to put it in other words: cognition follows *re*cognition from a developmental perspective on intersubjective relatedness. This opposes the portrayal of the solipsistic and monadic existence of the mind along the lines of which the idea of a primordial loneliness can become rejected. *Recognition* is, instead, primordial to all kinds of more objectifying modes of self-and-world disclosure, as also Axel Honneth (2018) has outlined in his theory of intersubjective ethical practice: Humans are always already affectively attuned to the world and engage with others in modes of interested participation. This forms an opposite to all more objectifying self-and-world disclosing modes. While a range of distortions of social relatedness might imply an "active" forgetting of the priority of the other (for instance, in a lack of empathic concern, as is the case in pathological

18 Heidegger has not further elaborated on the difference between solitude and loneliness and uses these as synonyms. We find such an almost interchangeable usage of 'aloneness', 'isolation', and also 'solitude' in many works on loneliness. One could bear in mind that "language […] has created the word 'loneliness' to express the pain of being alone. [A]nd it has created the word 'solitude' to the glory of being alone" (Tillich, 1963, p.17; chap.1). The difference between solitude and loneliness becomes most clear in the works of Hans-Georg Gadamer (1988), Hannah Arendt (1962), and Adrian Costache (2013), which I have discussed elsewhere.

narcissism; see: Jacobs, 2022) quite the opposite is revealed in the experience of loneliness: the lonesome soul is constantly reminded – it *remembers* – of what is essentially missing, namely, the relatedness to others stemming from being recognised as a participant in intersubjective processes of sense-making.

In the uncanny loneliness experience, the loner may have the feeling of becoming nearly invisible to or untouchable by others: Loneliness is the often harmfully experienced reminder that we all need basic recognition, and that this is what we normally are already taking for granted in our relations to the world and others. It is maybe not so difficult to understand that AI companions are designed in such a way that they exactly fill in this lacuna of an experienced lack of recognition. In my theory of loneliness, alienation is the structural (interpersonal) correlate to loneliness as a mere subjective feeling (on an intrapersonal level) and consequently leads to both individual suffering and social maladjustment. Agents normally can influence by which particular evaluative stances and particular modes of practice they relate to others, themselves, and the world. Normally, we experience oneself not as fundamentally exposed to or lost in social interaction, and we do not face severe distortions of "making sense" of the world, or experience our encounters with others as meaningless (even if we face sometimes difficulties or misunderstandings, etc.). This significantly changes in loneliness as alienation has a massive socially impairing dimension, which becomes evident, for instance, in the case of *Hikikomori*[19] in Japan. This yields a high vulnerability and risk to develop either a physical and/or mental illness that, in addition to the stigmatisation of loners (cf. Rotenberg and MacKie, 1999), point towards the bidirectional relationship of (psycho)pathology and loneliness on the one hand, and to loneliness as a social pathology, as discontent of digitalised cultures, on the other. This pathological dimension of loneliness will now be sketched.

2.3 The pathological dimension of loneliness

> In loneliness, the lonely one eats himself; in a crowd, the many eat him. Now choose!
> Friedrich Nietzsche, *Thus Spoke Zarathustra* (ZA 1883–1885; III)

---

19 "Hikikomori" is a term coined by Japanese psychologist Tamaki Saitō (cf. Social Withdrawal – Adolescence Without End (1998)) and refers to an "abnormal avoidance of social contact and literally translates as being confined". People withdraw themselves from society by seeking extreme levels of isolation – often staying in the same room for at least a period of 6 months – and show the regressive tendency, especially of males, to refuse leaving the house of their parents, to go to work, and instead to opt for staying isolated (except the contact with the hosting parents) in the same room for a period of at least 6 months.

Loneliness is the (often agonising) awareness of an inner distance from other people and the accompanying longing for connectedness in satisfying, sense-making relationships. The object of negative evaluation is the change for an agent's possibility for (participatory) sense-making. Nota bene: If this intersubjective dimension of "connectedness" and "experience of meaning" at the same time determines our ideas of the good life, then it (partly) defines our notions of biopsychosocial well-being, and loneliness can be perceived as some sort of impairment of well-being and health, respectively. It is characterised "as persistent state of emotional distress that occurs when persons feel alienated or misunderstood and rejected by others, and/or simply do not have the social contact to engage in the activities that give a person a sense of social inclusion and the opportunities for emotional intimacy" (Rook, 1984, p.1391), i.e., it includes a change in *affective resonance* and *emotional* experience (towards sadness, anger, envy, shame). Matthias Donath sees loneliness as "[...] in each personality individually and uniquely formed unison(s) of the most diverse emotional vibrations to a specific basic mood" (Donath, 1996, p.17; transl. KJA). Understood as a mood it is a broader affective background[20] against which specific emotions and emotional episodes develop in a specific situation. These affective experiences have a "biological background": Loneliness actually "hurts". Neurobiology associates it "there", where also physical pain is processed (anterior cingulate cortex – ACC; Eisenberger and Lieberman, 2004). The ACC is the cortical area that appears to be involved in the emotional reaction to pain rather than in the perception of pain itself (Price, 2000). Other neuroscientific studies suggest that in addition to its role in physical pain, the ACC is involved in monitoring painful social situations, such as exclusion or rejection, indicating that the ACC plays an important role in the detection and monitoring of social situations that cause social and/or emotional pain.[21] With respect to the *genetics* of loneliness it has been considered an adaptive response to isolation that gives the agent an impetus to try to re-integrate into social groups (Cacioppo and Patrick, 2008, p.201ff). The particular adaptive function of loneliness experiences may be that

20 This also makes loneliness classifiable in the category of the so-called existential feelings. Matthew Ratcliffe (2008) has described them as "background feelings", because they are normally not in the focus of our direct attention. Conceptualised as an affective notion ("mood-like"), loneliness is a specific disturbing experience when that sense of realness and reality that we normally always presuppose as self-evident modes of experience in our intentional processes of self-world-relatedness changes. Cf. Jacobs, 2013a; Jacobs et al., 2014.

21 Eisenberger et al. (2003) refer to their study of an fMRI virtual ball throwing game in which the ball was never thrown to one participant, the ACC showed activation, and this activation also correlated with self-reported measure of social distress.

"social" pain compels us to seek out social inclusion. In contrast, when associated with social withdrawal, loneliness may support regenerative processes (the felt need for being alone might also have a protective function). Depending on the resources available either empathic responses to the lonely are displayed (care) or active avoidance behaviour (Cacioppo and Patrick, 2008, pp.182–197), which is also explainable in terms of its affective *contagiousness* (Spitzer. 2018, pp.71–92). According to the interactive dynamics, the process of becoming lonely is a vicious circle in which agents get more and more lonely due to their being negatively biased[22], for instance, by focusing explicitly and exclusively on events of experienced social rejection, on their own (allegedly) negative traits, inabilities, and failures. Such biased negative self-attribution (Schultz and Moore, 1986) shows how faulty perceptions, irrational beliefs, etc. can deeply infiltrate lonely persons' self-understanding and world-orientation. Being in such a way attuned to the world, the lonesome find themselves evidenced by every single negative experience and may often also tend to misinterpret situations as explicit forms of social neglection, exclusion, and rejection. This is evidenced by social network theories on loneliness, which emphasise that not only difficulties in trusting peers (cf. Eberhart and Hammen, 2006) but also lack of support (cf. Kerr et al., 2006) and experiences of social rejection by others (Cassidy and Asher, 1992) are leading to loneliness and also to depression (Heinrich and Gullone, 2006; Nangle et al., 2003; Vanhalst et al., 2012). Besides vulnerability and resilience factors[23], especially reinforcement by the environment influences these self-priming looping dynamics of a "vicious circle of being a social outcast" (Cacioppo and Hawkley, 2005). This transformation towards becoming an "outsider" is further evidenced in studies on loneliness typical behaviours: Lonely people are more inclined to trust others less (e.g., in paraphrenia cf. Fuchs, 1999; and Janzarik, 1973), to negatively perceive others (e.g., as fierce), and to approach social interactions in a more hostile, or defensive, manner (cf.; Cacioppo and Hawkley, 2009). Being among others apparently can be a highly ambivalent experience and may trigger the desire rather to continue avoiding social contact or: to prefer an AI companion instead of dealing with potentially stressing social situations with other humans. In order to crack the negative feedback loops that reinforce the feelings of loneliness, in severe cases specific addi-

22 Generally, a bias is defined as a proclivity to take one direction over another which under same conditions will lead to accuracy or realism, but under other conditions will lead to inaccuracy, while "distortion" implies something invariably wrong (cf. Power and Dalgleish, 2008).

23 Loneliness is evaluated as either negative (ego-dystonic) or positive (ego-syntonic) experience, but studies show that lonely people, even if they have arranged themselves with being chronically lonely, still have intense social needs (Shaver et al., 1985). For cross-cultural variations of coping see: Rokach et al., 2000.

tional therapeutic inventions are needed to re-shape these perpetuating dynamics. As already mentioned, we find also profitable interaction dynamics in the sense of health-promotion in the coupling of AI and humans. The latter is the case, for example, when people learn to assess and evaluate things more realistically through computer simulations (e.g. in the case of phobias) or experience their physicality anew through the use of therapeutic AI, or can even try to "overwrite" their unconscious fears with the help of "decoded neurofeedback" that aims to reduce fear, without triggering the actual fear memory by using artificial intelligence algorithms (Koizumi et al., 2017). Especially the latter appears as an intriguing technique to intervene with the unconscious levels of loneliness, but this technique is nothing an AI companion (to date) can provide (it therefore would have to be equipped with a brain scan). Although some surface symptoms of loneliness can become "eased" (e.g., by providing the possibility for sexual intimacy, as love dolls like *Harmony* promise), *the core condition of loneliness* cannot be altered by companion AI relatedness alone, especially not when the companion robot "echoes" the loner's perspective, which is, as we have seen, often a strongly biased one that is rather oriented towards even further social withdrawal and avoidance. The recent companion AI seems to me rather a "quick fix" that allows only a temporal forgetting, but never a real escape from the discontent of loneliness. One can rather assume that someone may need therapy exactly for this reason: for explicitly favouring an x-bot over a human, i.e., favouring the monologic over dialogic existence, under the assumption that companionship with another human is not totally out of experiential reach. But is there really nothing more conciliatory to say? Maybe in the ways the parallel to the role *of transitional objects*[24] (Habermas, 2020) can be drawn here. These objects are not called so for nothing, because precisely they must become left behind in order to find oneself, to mature and individuate. AI companions may be these beloved objects the lonely have to leave behind, in order to really tackle their loneliness, but which nevertheless can be kept in good memory and cherished as a temporary concomitant of one's loneliness. Consequently, I suggest that especially those treatment forms can be emphasised for a good transition, which particularly aim to help agents to develop social support networks (Peplau and Perlman, 1982; Young, 1982) so that it is with rather therapeutic intervention (with fellow-beings) that loneliness can become revealed in its

---

24 The term was coined by Donald Winnicott (1953) and refers to a first object (a possession such as a teddy bear) that a child feels attached to and emotionally affected towards and which is recognised as something separate from him or herself. They originate in that developmental phase when inner and outer reality begin to become apparent, so these objects are "not-me" and "me" at once. They take the place of the mother-child bond and facilitate the transition from the omnipotence stage to the capacity to "objectively perceived" relations. See also: Winnicott (1971).

full uncanniness, worked through to open up new experiential horizons of recognition.

The social-psychological reference category of recognition allows to finally add a culture-analytic perspective on loneliness as a discontent of and in digitalised cultures. This completes a view on the pathological dimension of loneliness in stressing the *social pathological dimension* of loneliness: Freud describes in his drive-dualistic theory of culture (Freud, 1930) not only the sufferings of neurotic or psychotic individuals, which are explained by an overstraining repressive sexual ethos, but also its culture-threatening potential. Cultural processes generally have the function of ensuring survival (through the peaceful use of human aggressive energy) and making coexistence conducive (primarily through civilisational progress using libidinous energy, i. e., in sublimation). Thus, culture serves to protect man against nature in the trajectories of normative regulation of interpersonal relations (Freud, 1930, p.449). Cultivation through technology and knowledge expands our possibilities to dominate nature, particularly where it frustrates our desire for happiness, while the cultural shaping of social relations is the basis for everything else (Jacobs and Kettner, 2016). The companion robot is a technical masterpiece attempting to counteract loneliness – this peculiar discomfort in today's cultures that explicitly revels in that very digital connectedness – and it is undoubtedly a new form of cultural design of social relationships. But is it really reconciling enough for the suffering from the cultural performance? A second insight we gain from Freud is that a reduction in the ability of agents to love and work becomes clinically conspicuous, and that the clinical side of discontent in culture shows itself where culture production can only be maintained with a huge expenditure of energy (Freud, 1908). Loneliness may be caused by forms of drive renunciation that are culturally imposed, but that not all people can deal with equally well. The genesis and dynamics of both individual pathology and cultural pathology are explained in terms of repression and sublimation, but the latter can only fulfil its culture-promoting and stabilising function as long as it allows for drive satisfaction, besides other gratifications, so that individuals can tolerate the drive expenditure necessary for cultural adaptation. The phenomena of increasing loneliness and/or illness (which have turned out to strongly correlate) could be seen as a paradigmatic expression of not succeeding well. Loneliness is clinically conspicuous in terms of individual pathology *and* at the same time qualifies as a socio-cultural pathology, if one agrees on the fact that people "fall ill" with loneliness precisely because they (have to) strive to meet the dominant cultural norms, which is a "struggle for recognition" (Honneth, 2005). As far as the "goals" of cultural norming are concerned one should therefore focus on the social role-models (stereotypes) that are associated with great social recognition, professional success, and happiness in a given socio-cultural context, in order to include an analysis of the specific material background conditions (in socioeconomic, and not exclusively drive-economic, terms) that negatively impact well-being and health in recent societies and are leading to increased loneliness rates worldwide.[25] This would be a matter of future study on loneliness experiences in its bidirectional relation with social exclusion against the backdrop of an analysis of precarious living conditions (such as poverty, bad working conditions, reduced access to health care, etc.).

To sum it up: Not only can seeking ease in an AI relationship be reinterpreted as a compromise formation on the intrapsychic level, but also from a cultural-analytic point of view, while the intrapsychic conflict of loneliness leads to a range of challenging experiences of a changed sense for self, others, and the world, on the interpersonal level it shows as alienation and (mal)adaptive praxis. As a social pathology, loneliness is the uncanny recall of the autonomous agent in its (digital) struggles for social recognition.

3 Conclusion

AI companionship may not be considered in the literal sense the philosopher's stone in the fight against loneliness, but it can be conciliatory to add that social AI applications in the form of companion x-bots may improve in the future, and that the therapeutic possibilities for alleviating the symptoms of loneliness with it are certainly only just beginning.

This chapter emphasised four things. First, we as relational beings necessarily need social recognition. Where there is an experienced lack of it, e.g., because one does not get along so well in the struggles for recognition and is even exposed to forms of disrespect and reification, alienation can develop, which conceptually represents the social-psychological and culture-reflexive correlate to the category of individual-psychological loneliness experiences of changed sense-making, thus affectivity and cognition. I have outlined that loneliness can be accompanied by a significant change in experiential possibilities, and that this often includes a change or even loss of meaningful relatedness to self, others, and the world. Thirdly, I hope I have given sufficient reasons to accept that AI companionship cannot change the very situation of loneliness, but rather might be perceived as a "cultural trap". This overall points to a possible weakening rather than a strengthening of

---

25 See therefore the WHO report on loneliness included in the WHO strategies for Healthy Ageing: https://www.who.int/teams/social-determinants-of-health/demographic-change-and-healthy-ageing/social-isolation-and-loneliness (accessed 28.11.2022). See also: Keating (2022).

agent autonomy in trying to exclusively cope with loneliness in an exclusive digital way. I have argued that digital loneliness as a new form of cultural discontent paradigmatically shows in the attempt to compensate for a lack of social recognition and that I suspect that these struggles for recognition can be satisfactorily solved with AI companion. Fourthly, I have outlined a few differences that seem to be decisive when we think about humanising of AI: x-bots will perhaps accompany us more and more in the future (e. g., as transitional objects), and in this respect, nothing human is alien, simply because these canny objects mirror us. Humanising AI may include as a future idea that x-bots don't stay alien to us forever, but perhaps they *should* remain it, at least, if we don't want to forget ourselves in a soliloquy that misses the call of the uncanny for recognition.

References

AltHumans. (n.d.). *RealDoll Companion Robot Store.* AltHumans. https://www.althumans.com/companion-robots/real-doll.html (accessed 24. 11. 2022).

Arendt, H. (1962). *The Origins of Totalitarianism.* Meridian Books.

Neves, B. B., Waycott, J., & Maddox, A. (2021). When Technologies are Not Enough: The Challenges of Digital Interventions to Address Loneliness in Later Life. *Sociological Research Online, 28*(1), pp. 150–170.

Breazeal, S., & Scassellati, B., (1999). How to build a robot that makes friends and influence people. *IEEE/RSJ International Conference on Intelligent Robots and Systems (IROS), 2*, pp. 858–863.

Cacioppo, J. T., & Hawkley, L. C. (2005). People Thinking About People: The Vicious Cycle of Being a Social Outcast in One's Own Mind. In K. D. Williams, J. P. Forgas, & W. von Hippel (Eds.), *The social outcast: Ostracism, social exclusion, rejection, and bullying*, pp. 91–108. Psychology Press.

Cacioppo, J. T., & Hawkley, L. C. (2009). Perceived social isolation and cognition. *Trends in cognitive sciences, 13*(10), pp.447–454.

Cacioppo, J. T., & Patrick, W. (2008). *Loneliness: Human nature and the need for social connection.* W. W. Norton & Co.

Cassidy, J., & Asher, S. R. (1992). Loneliness and Peer Relations in Young Children. *Child Development, 63*(2), pp.350–365.

Clark, D. A., Beck, A. T., & Alford, B. A. (1999). *Scientific foundations of cognitive theory and therapy of depression.* John Wiley & Sons Inc.

Colombetti, G. (2014). *The Feeling Body: Affective Science Meets the Enactive Mind.* Cambridge/MA: MIT Press.

Costache, A. (2013). On solitude and loneliness in hermeneutical philosophy. *Meta: Research in Hermeneutics, Phenomenology, and Practical Philosophy, 5*(1), pp.130–149.

De Jaegher, H., Di Paolo, E., & Gallagher, S. (2010). Can social interaction constitute social cognition?. *Trends in cognitive sciences, 14*(10), pp.441–447.

Di Paolo, E., & De Jaegher, H. (2012). The interactive brain hypothesis. *Frontiers in human neuroscience, 6*, p.163.

Di Paolo, E. A., Rhode, M., & De Jaegher, H. (2010). Horizons for the enactive mind: Values, social interaction, and play. In Enaction: Towards a New Paradigm for Cognitive Science. In J. Stewart, O. Gapenne, & E. Di Paolo (Eds.), pp. 33–87. MIT Press.

Donath, M. (1996). Begreifen, Bewerten, Behandeln von Einsamkeit. In B.M-v. Meibom (Ed.), *Einsamkeit in der Mediengesellschaft*, (pp.15–32). Lit-Verlag.

Eisenberger, N. I., & Lieberman, M. D. (2004). Why rejection hurts: a common neural alarm system for physical and social pain. *Trends in cognitive sciences*, *8*(7), pp.294–300.

Eisenberger, N. I., Lieberman, M. D., & Williams, K. D. (2003). Does rejection hurt? An FMRI study of social exclusion. *Science (New York, N.Y.)*, *302*(5643), pp.290–292.

Franzen, J. (2002). How to Be Alone. New York: Farrar, Straus and Giroux.

Freud, S. (1908). *Die "kulturelle" Sexual-moral und die moderne Nervosität.* GW VII, S. 141–167 [Trans.: '"Civilized" Sexual Morality and Modern Nervousness' 1924 C.P., 2, 76–99. In *The Standard Edition of the Complete Psychological Works of Sigmund Freud.* ed. Strachey, J. 1955, Macmillan].

Freud, S. (1909). *Bemerkungen über einen Fall von Zwangsneurose'*, G.S., 8, 269; G.W., 7, 381. (124, 319) [Trans.: 'Notes on a Case of Obsessional Neurosis', C.P., 3, 293. In J. Strachey (Ed.) *The Standard Edition of the Complete Psychological Works of Sigmund Freud.* 1955, pp.10:i–vi, 153, Macmillan].

Freud, S. (1911). *Formulierungen über die zwei Prinzipien des psychischen Geschehens*, G.S., 5, 409; G.W., 8, 230. (246) [Trans.: 'Formulations on the Two Principles of Mental Functioning', C.P., 4, 13. In J. Strachey (Ed.) *The Standard Edition of the Complete Psychological Works of Sigmund Freud.* 1955, pp. 13–21, 13, Macmillan].

Freud, S. (1920). *Jenseits des Lustprinzips*, Vienna. G.S., 6, 191; G.W., 13, 3. (140) [Trans.: 'Beyond the Pleasure Principle'. In J. Strachey (Ed.) *The Standard Edition of the Complete Psychological Works of Sigmund Freud.* 1955, p.18, 7, Macmillan].

Freud, S. (1926). *Hemmung, Symptom und Angst*, Vienna. G.S., 11, 23; G.W., 14, 113. (116, 141, 196, 236, 319–20) [Trans.: 'Inhibitions, Symptoms and Anxiety', London, 1936; 'The Problem of Anxiety', New York, 1936. In J. Strachey (Ed.) *The Standard Edition of the Complete Psychological Works of Sigmund Freud.* 1955, Macmillan].

Freud, S. (1930). *Das Unbehagen in der Kultur*, Vienna. G.S., 12, 29; G.W., 14, 421. (248) [Trans.: 'Civilization and its Discontents', London and New York, 1930. In J. Strachey (Ed.) *The Standard Edition of the Complete Psychological Works of Sigmund Freud.* 1955, p.21, Macmillan.]

Friedman, C. (2020). Human-Robot Moral Relations: Human Interactants as Moral Patients of Their Own Agential Moral Actions Towards Robots, In A. Gerber (Ed.) *Artificial Intelligence Research. SACAIR 2021. Communications in Computer and Information Science*, 1342. Springer, Cham.

Fuchs, T. (1999). Patterns of relation and premorbid personality in late paraphrenia and depression. *Psychopathology*, *32*(2), pp.70–80.

Fuchs, T. (2011). The brain–A mediating organ. *Journal of Consciousness Studies 18*(7–8), pp.196–221.

Fuchs, T., & De Jaegher, H. (2009). Enactive Intersubjectivity. Participatory Sense-Making and Mutual Incorporation. *Phenomenology and the Cognitive Sciences*, *8*, pp.465–486.

Gadamer, H. G. (1988). Isolation as a Symptom of Self-Alienation. In Praise of Theory. Speeches and essays, pp.101–113, Yale UP.

Gallotti, M. L. (2013). Social cognition in the we-mode. *Trends in Cognitive Sciences*, *17*(4) pp.1–6.

Gao, J., Galley, M., & Li, L. (2018). Neural Approaches to Conversational AI. *The 41st International ACM SIGIR Conference on Research & Development.* arXiv:1809.08267

Geniusas, S. (2012). *The origins of the horizon in Husserl's phenomenology.* Springer.

Habermas, T. (2020). *Geliebte Objekte: Symbole und Instrumente der Identitätsbildung* (Vol. 19). Walter de Gruyter GmbH & Co KG.

Heidegger, M. (1962). *Being and Time* (J. Macquarrie & E. Robinson, Trans.). Harper & Row.

Heinrich, L. M., & Gullone, E. (2006). The clinical significance of loneliness: A literature review. *Clinical Psychology Review 26*(6), pp.695–718.

Honneth, A. (2018). Reification and Recognition. In M. Jay (Ed.), *Reification: A New Look at an Old Idea*, (pp.17–94, 40–52). Oxford UP.

Honneth, A. (2005). *The Struggle for Recognition; The Moral Grammar of Social Conflicts* (J. Anderson, Trans.). Polity Press. (Original work published 1992)

Husserl, E. (2008). *Die Lebenswelt: Auslegungen der Vorgegebenen Welt und Ihrer Konstitution. Texte aus dem Nachlass (1916–1937).* Springer Verlag.

Jacobs, K. A. (2013a). The depressive situation. *Frontiers in Theoretical and Practical Psychology, 4*(429), pp.1–10.

Jacobs, K. A. (2013b). Loneliness in Philosophy, Psychology, and Literature by Ben Lazare Mijuskovic, iUniverse 2012. *Metapsychology Online, 17*(39).

Jacobs, K. A. (2022). The concept of Narcissistic Personality Disorder–Three levels of analysis for interdisciplinary integration. *Frontiers in Psychiatry, 16* (Sec. Personality Disorders).

Jacobs, K. A., Stephan, A., Paskaleva-Yankova, A., and Wilutzky, W. (2014). Existential and Atmospheric feelings in Depressive Comportment. *Philosophy, Psychiatry & Psychology 21*(2) pp.89–110.

Jacobs, K. A., and Kettner, M. (2016). Zur Theorie "sozialer Pathologien" bei Freud, Fromm, Habermas und Honneth. In M. Clemenz, H. Zitko, M. Büchsel, and D. Pflichthofer (Eds.) *IMAGO. Interdisziplinäres Jahrbuch für Psychoanalyse und Ästhetik, 4,* 119–146.

James, W. (1884). What is an Emotion?. Mind, 9, pp.188–205.

Janzarik, W. (1973). Über das Kontaktmangelparanoid des höheren Alters und den Syndromcharakter schizophrenen Krankseins. *Der Nervenarzt, 44*(10), pp.515–526.

Johnston, C. (2022). Ethical Design and Use of Robotic Care of the Elderly. *Bioethical Inquiry, 19*(1), pp.11–14.

Johnston, L. (2018). Japanese Man Marries Hatsune Miku Gatebox Device. *OTAQUEST.* https://www.otaquest.com/hatsune-miku-gatebox-marriage/ (accessed 25.11.2022)

Jorba, M. (2020). Husserlian horizons, cognitive affordances and motivating reasons for action. *Phenomenology and the Cognitive Sciences*, 19, pp.847–868.

Keating, N. (2022). A research framework for the United Nations Decade of Healthy Ageing (2021–2030). *European Journal of Ageing, 19* pp.775–787.

Kierkegaard, S. (2012). *Der Begriff Angst/Die Krankheit zum Tode.* Göttingen.

[Trans.: 'Søren Kierkegaard 2013. Kierkegaard's Writings', XIX, Volume 19: *Sickness Unto Death: A Christian Psychological Exposition for Upbuilding and Awakening* (Vol. 86). Princeton University Press].

Koizumi, A., Amano, K., Cortese, A., Shibata, K., Yoshida, W., Seymour, B., Kawato, M., & Lau, H. (2016). Fear reduction without fear through reinforcement of neural activity that bypasses conscious exposure. *Nature human behaviour, 1,* 0006.

Langlitz, N. (2005). *Die Zeit der Psychoanalyse. Lacan und das Problem der Sitzungsdauer.* Suhrkamp.

Laplanche, J., & Pontalis, J. B. (1988). The Language of Psychoanalysis. Routledge.

Lefèvre, E. (1986). Ich bin ein Mensch, nichts Menschliches ist mir fremd. In O. Herding (Ed.) *Wegweisende Antike: zur Aktualität humanistischer Bildung; Festgabe für Günter Wöhrle.* Stuttgart: Württemberg. Verein zur Förderung d. Humanist. Bildung. 1, pp.39–49.

McLuhan, M. (2017). *The Lost Tetrads of Marshall McLuhan.* OR Books.

Mijuskovic, B. L. (2012). *Loneliness in Philosophy, Psychology, and Literature.* Bloomington iUniverse.

Mijuskovic, B. L. (2015). *Feeling Lonesome. The Philosophy and Psychology of Loneliness.* Praeger.

Nangle, D. E., Erdley, C. E., Newman, J. E., Mason, C., & Carpenter, E. M. (2003). Popularity, friendship quantity, and friendship quality: Interactive influences on children's loneliness and depression. *Journal of Clinical Child and Adolescent Psychology, 32* pp.546–555.

Nietzsche, F. (2021). Also Sprach Zarathustra. III. In G. Colli & M. Montinari (Eds.) *Band 4 Also sprach Zarathustra I–IV. Kritische Studienausgabe*, (pp. 191–292). De Gruyter.

Kohut, H. (1978). *The search for the self: Selected writings of Heinz Kohut: 1950–1978 (Vol. 2).* International Universities Press.

Papoušek, H. & Papoušek, M. (1992). Beyond emotional bonding: The role of preverbal communication in mental growth and health. *Infant Mental Health Journal, 13*, pp.43–53.

Peplau, L. A., & Perlman, D. (1982). Perspectives on Loneliness. In L.A. Peplau and D. Perlman (Eds.) *Loneliness: A sourcebook of current theory, research, and therapy* (pp.1–18). Wiley.

Ulrich, J.P., Timmermans, B., Vogeley, K., Frith, C. D., & Schilbach, L. (2013). Towards a neuroscience of social interaction. *Frontiers in Human Neuroscience, 7.*

Possati, L. (2021). *The Algorithmic Unconscious. How Psychoanalysis Helps in Understanding AI.* Routledge.

Power, M. J., & Dalgleish, T. (2008). *Cognition and emotion. From Order to Disorder.* Psychology Press.

Price, D. D. (2000). Psychological and neural mechanisms of the affective dimension of pain. *Science, 288*(5472), pp.1769–72.

Purington, A., Taft, J. G., Sannon, S., Bazarova, N. N., and Taylor, S. H. (2017). "Alexa is my new BFF" Social Roles, User Satisfaction, and Personification of the Amazon Echo. In *Proceedings of the 2017 CHI conference extended abstracts on human factors in computing systems*, pp. 2853–2859.

Ratcliffe, M. (2008). *Feelings of being. Phenomenology, Psychiatry and the Sense of* Reality. Oxford UP.

Rokach, A., Bacanli, H., & Ramberan, G. (2000). Coping with Loneliness: A Cross-Cultural Comparison. *European Psychologist, 5*(4), pp.302–311.

Rook, K. S. (1984). Promoting social bonding: Strategies for helping the lonely and socially isolated. *American Psychologist, 39*(12), pp.1389–1407.

Rotenberg, K. J. & MacKie, J. (1999). Stigmatization of social and intimacy loneliness. *Psychological Reports, 84*(1), pp.147–8.

Schilbach, L., Timmermans, B., Reddy, V., Costall, A., Bente, G., Schlicht, T., & Vogeley, K. (2013). Toward a second-person neuroscience. *Behavioral and Brain Sciences, 36*(4), pp.393–414.

Shaver, P., Furman, W., and Buhrmester, D. (1985). Aspects of a life in transition: Network changes, social skills and loneliness. In S.W. Duck & D. Perlman (Eds.) *Understanding personal relationships*, (pp.193–219). Sage.

Schultz, N. & Moore, D. (1986). The Loneliness Experience of College Students: Sex Differences. *Personality and Social Psychology Bulletin, 12*, pp.111–119

Skjuve, M., Følstad, A., Fostervold, K. I., Brandtzaeg, P. B. (2021). My Chatbot Companion – a Study of Human-Chatbot Relationships. *International Journal of Human-Computer Studies, 149*, pp.1–14.

Spitzer, M. (2018). *Einsamkeit-die unerkannte Krankheit: schmerzhaft, ansteckend, tödlich.* Droemer.

Szollosy, M. (2017). Freud, Frankenstein and our fear of robots: projection in our cultural perception of technology. *AI & Society, 32*, pp.433–439.

Tamaki, S. (2013). *Hikikomori: Adolescence without End* (J. Angles, Trans.). University of Minnesota Press.

Thompson, E. (2007). *Mind in Life: Biology, Phenomenology, and the Sciences of Mind.* Harvard University Press.

Thompson, E., & Stapleton, M. (2009). Making sense of sense-making: Reflections on enactive and extended mind theories. *Topoi, 28*(1), pp.23–30.

Tillich, P. (1963). *The Eternal Now.* Scribners.

Turkle, S. (2004). Wither psychoanalysis in computer culture? *Psychoanalytic Psychology, 21*(1), pp.16–30.

Turkle, S., Taggart, W., Kidd, C. D., & Dasté, O. (2006). Relational artifacts with children and elders: the complexities of cypercompanionship. *Connection Science, 18*(4), pp.347–361.

Vanhalst, J., Klimstra, T. A., Luyckx, K., Scholte, R. H. J., Engels, R. C. M. E., & Goossens, L. (2012). The interplay of loneliness and depressive symptoms across adolescence: Exploring the role of personality traits. *Journal of Youth and Adolescence, 41*(6), pp.776–787.

World Health Organization (WHO). (2000). World Report on Ageing and Health. https://www.who.int/teams/social-determinants-of-health/demographic-change-and-healthy-ageing/social-isolation-and-loneliness (accessed 28.11.2022).

Winnicott, D. (1953). Transitional objects and transitional phenomena. *International Journal of Psychoanalysis, 34*, pp.89–97.

Winnicott, D. (1971). *The use of an object and relating through identifications. In Playing and reality.* Tavistock. pp. 101–121.

Young, J. E. (1982). Loneliness, depression and cognitive therapy: theory application. In L.A. Peplau & D. Perlman (Eds.) *Loneliness: A sourcebook of current theory, research and therapy,* (pp.1–18). Wiley.

Andre Nusselder

How football became posthuman: AI between fairness and self-control

Abstract: Decision-making by football referees is increasingly supported by Video Assistance Referee (VAR) technologies, with the goal of implementing AI so that accurate and fair decisions can be made without interrupting the flow of the game too much. This chapter analyses the connection between these technologies and affective self-regulation of those involved. It does so from Norbert Elias'theory of civilisation, in which he analyses – using Freud's metapsychology – how increased civilised behaviour leads to increased self-control of individuals. The chapter argues that the aim of making football a fairer game with the use of AI has a similar impact on those involved and is thus a next step in the movement towards the posthuman condition which takes place subtle as humans adapt to it without generally being conscious of it but with far-reaching effects. Herein intelligent technologies and human behaviour are increasingly intertwined, with a different and more restrictive tuning of the individual's libidinal economy as a result.

Keywords: metapsychology, AI, football, posthuman

Introduction

This chapter analyses AI in professional football. By departing from Elias's theory of civilisation it holds that AI contributes to the interdependencies in these playful activities and changes the behaviour of those involved. It studies the transformation of the human person in the age of intelligent technologies by means of a case study of decision-making in the game of football.

> The more the strong contrasts of individual conduct are tempered, the more the violent fluctuations of pleasure or displeasure are contained, moderated and changed by self-control, the greater becomes the sensitivity to shades or nuances of conduct, the more finely attuned people grow to minute gestures and forms, and the more complex becomes their experience of themselves and their world at levels which were previously hidden from consciousness through the veil of strong affects. (Elias, 2000, p. 418–419)

André Nusselder is a Dutch philosopher and author, who works as Academic director and senior researcher Education & Technology at Marnix Academie Utrecht

https://doi.org/10.1515/9783111007564-005

1 Sports and (AI) technologies

The value of sports

As a Dutchman and football enthusiast one of the traumatic events inscribed into my life's (pre)history is the lost World Cup final in 1974 of the Dutch national team against Germany. In the 25th minute, when the score was 1:0 in favour of the Dutch team, German striker Bernd Hoelzenbein received the ball and dribbled towards the Dutch goal where, while entering the penalty area, he was tackled by Dutch midfielder Wim Jansen and fell down. The German team was awarded a penalty kick by the English referee Jack Taylor, which was used by Paul Breitner, and the Germans eventually won the game by 2:1. The referee's decision was crucial in the game and remains controversial until today, as Hoelzenbein was accused of being dishonest because of faking and exaggerating the collision, which he always vehemently denied (Sabag et al, 2018; Tamir and Bar-eli 2021). The example shows the difficult issue of human decision-making in a dynamic context such as a football game.

Football is an activity with the goal of providing entertainment or amusement, for players and spectators, a competitive activity or sport in which players contend with each other according to a set of rules. The activity is guided by intrinsic values such as pleasure and health and by the extrinsic value of fair play: playing according to the rules and not having an unfair advantage. However, not all perspectives on sports have this value-driven approach in it (see Loland 2002). Sports can be, first of all, an activity focused on performance and success, where history has shown that all sorts of means (and even prohibited substances such as doping in professional cycling) are used to achieve these goals. Secondly, sports can also be a means to achieve goals of political, ideological, or financial prestige – like promoting the importance and significance of a country,[1] or gaining prestige by investing in football clubs.[2] The activity is instrumental to achieving external goals

---

1 As formulated in the abstract of a paper on AI: "In order to respond to the call of the country to build a strong sports country, the workers in the sports industry should speed up the pace of sports development. Artificial intelligence is a high-end industry, but also a key technology to guide the construction of sports power" (Zhang and Li 2021).

2 This is an increasingly popular activity of oligarchs or oil-billionaires: Tamim bin Hamad Al Thani, the Emir of Qatar, has owned Paris Saint-Germain since 2011 through state-run shareholding organisation Qatar Sports Investments; Sheik Mansour owns Manchester City through the British-based holding company City Football Group, with the majority stake owned by the Abu Dhabi United Group.

(prestige, performance). In the third, value-driven perspective the activity of sports has a value in itself that must be accommodated and guaranteed as much as possible for all the participants. Sports then is an arena for human development guided by important virtues such as equality and fairness: the Olympic Games as a gathering of human potential where each individual can show and enjoy their talents in common activities guided by the principle that 'participating is more important than winning' ("L'important c'est de participer") as formulated by the founder of the modern Olympic Games, the French educator and historian Pierre de Coubertin (1863–1937). Although fair play is then the supreme and guiding value of the game, this value is not always met – and this is where technology is brought in as a help.

The VAR in football

The involvement of technology in sports can be divided into technologies that help athletes to achieve better performances and technologies that are used as a governing mechanism of sport. In the second category, many of the recent innovations in sports focus on the aim of assisting the referees and the regulatory process (Tamir and Bar-eli 2021, 4). In rugby, cricket, American football, tennis, and field and ice hockey technology and video evidence are routinely used to support match officials. Football hesitated to employ technologies and only introduced the goal-line technology in 2012 in order to precisely determine whether the ball has fully crossed the line (if the ball only touches the line for the tiniest bit, it is not a goal according to the Laws of the Game). However, the linkage to technology significantly changed with the introduction of the Video Assistance Referee (VAR), an example of autonomous assistance given to soccer referees. The VAR was tested for the first time during the 2012–2013 season but was only officially introduced into the Laws of the Game in 2018 to help referees in reviewing decisions made by the head referee by means of video footage, and the IFAB (International Football Association Board) considers it 'a historic step for greater fairness in football.'[3] According to IFAB Principles, a VAR is a match official, with independent access to match footage, with the task of assisting the referee by immediately monitoring potentially match-changing incidents, which are events of 'clear and obvious error' or 'serious missed incident' in relation to: (a) Goal/no goal; (b) penalty/no

3 https://www.theifab.com/news/historic-step-for-greater-fairness-in-football/ (accessed March 24, 2023).

penalty; (c) direct red card (not second yellow card/caution); and (d) mistaken identity (when the referee cautions or sends off the wrong player of the offending team). In case the video evidence shows the initial decision of the referee to be erroneous, VAR recommends an on-field review (OFR), except in cases of VAR-only decisions, which are factual decisions concerning onside/offside[4], whether the ball is in/out of play, and whether a foul occurred inside/outside of the penalty area. In case of an on-field review the referee is called to a monitor next to the pitch to review the incidents from the video evidence that is available (usually from different angles available to, and deemed significant by, the VAR-team). The final decision is always taken by the referee, either based on the video footage from the field review or based on the information from the VAR-team.[5]

Where previously the criticism concerned above all the imperfections of the 'situated decisions' of the referee on the pitch, with the significant role of technologies also another strand of criticism appeared. The VAR would disrupt the flow and pace of the game due to the interruption and long time taken for reviewing the video footage, which would cause viewers to lose interest. Also, players' activities were interrupted, and spectators would not be able to fully cheer a goal as it is always followed by a time for the VAR to check whether no possible faults were committed in the build-up to the goal: celebrating it to the full can only take place after this delay and time of uncertainty. Soccer executives, such as former FIFA chairman Sepp Blatter, also raised many concerns about technologies undermining the authenticity of the game and thus expressed a longer tradition of opposition to new technologies as threatening sports traditions and communities (see Gelberg 1998). Furthermore, also with the VAR inaccurate decisions occur as a result of human errors. In order to overcome these imperfections, VAR technologies are currently supplemented with artificial intelligence. Due to the visual nature of on-/offside decisions, especially computer vision techniques seem promising.

AI and sports

Artificial intelligence is broadly understood as intelligence demonstrated by machines by means of intelligent agents or computer systems that perceive their environment for achieving certain goals. As it is a very broad field, and the definition

4 Whether another player than the player receiving the ball was in offside position and with that gave advantage to the player receiving the ball is – as will later on be explained – not such a clear black-or-white decision but a discretionary decision with a high degree of interpretation.
5 VAR has proven to be much more accurate than the calls of a human referee, and the accuracy in decisions is supposed to have increased from 92.1% to 98.3% (Spitz et al, 2021).

of the term intelligence is furthermore ambiguous, it is difficult to provide a uniform definition. Nevertheless, artificial intelligence is about the characteristics of IT systems to exhibit human-like, intelligent behaviour by means of four core abilities: perception, understanding, action, and learning. The foundations of artificial intelligence consist of big data (large and complex data sets), cloud computing (network access to data sets, networks, servers, storage, applications, and services), Internet of Things (exchanging data gained from objects that have sensors or software embedded in them over the Internet or other communication channels), and algorithms (formal steps for processing the data).

Artificial intelligence was founded as an academic discipline in the 1950s by developing instructions for the machinic processing of information so that, for instance, a computer could develop a checker strategy. The next big step was the development of machine learning (ML) in the 1980s, where not only does the processing of information occur according to preestablished programs, but also information from past experiences is used to learn so that complex problems can be solved. In supervised ML the program is provided example data and interpretations in advance in order to provide predictions (as in image recognition). In unsupervised ML the computer program must recognise structures without such an assignment or 'lesson' of the example data (for instance, recognising patterns in user behaviour). In reinforcement ML the program interacts with its environment and remembers the consequences of its actions (for example, Google's Deep Mind). The third milestone is deep learning as a further development of ML and is pivotal to understanding AI today. In DL the program can teach itself to learn through thousands of examples, and this learning occurs in 'brain-like' networks of artificial neurons that map an artificial neural network (ANN) that consist of patterns and forecasts so that a specific complex problem can be solved (Gottschalk, Tewes & Niestroj, 2020, p. 39)

In sports, the use of artificial intelligence is first and foremost deployed for using data of athletes for improving their performance and/or enhancing team competitiveness (training). It is also deployed for improving customer experience (such as providing television viewers with real-time data of the players or game), or for predicting sporting outcomes (Rathi et al 2020; Huang 2020; Pretorius& Parry, 2016). For the present chapter, the use of artificial intelligence in sports will be narrowed to computer vision: a field of artificial intelligence (AI) that enables computers and information systems to derive meaningful information from digital images, videos, and other visual inputs – and take actions or make recommendations based on that information. In sports, computer vision is used for several purposes, such as improving the broadcast viewer experience: in ice hockey there is the hockey puck tracking system that helps to view the (difficult to see) hockey puck during a match, and in (American) football and rugby there is the pos-

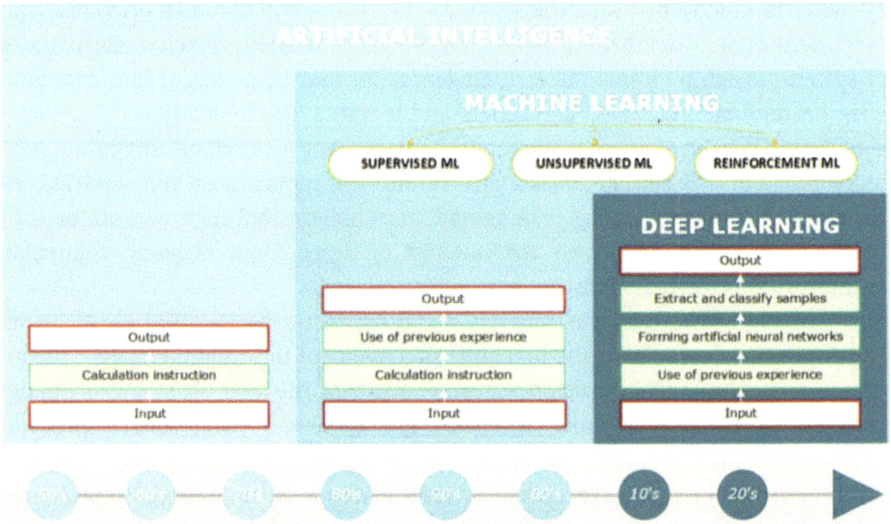

Figure 1: How machine learning and deep learning systems work. Source: in Gottschalk, Tewes & Niestroj (2020, p. 40)

sibility for drawing virtual marks across a football field. Computer vision also allows for improving the training process of professional athletes (for instance, manually fitting a skeleton model upon the body of a long jumper in order to estimate the centre of mass), and automatic sports analysis and interpretation (as the tracking of players in a sports game), or for commercial benefit (such as real-time billboard substitution in video streams). This chapter focuses on the use of computer vision to improve referee decisions: the VAR is an excellent case for studying human judgement and decision-making.[6]

AI and decision-making in football

At Wimbledon, the hawk-eye system Watson (IBM) is used for deciding whether the ball has touched the ground in – or outside – the playing field. In football, Goal-Line Technology has gained a high level of acceptance as the technology only interferes in black-or-white decisions of right or wrong: by monitoring the boundary lines with photoelectric sensors it is clear whether the ball has crossed the line or not. Who then earns the throw-in, kick-off, or corner kick remains, how-

6 See Bar-eli (2011), Raab et al. (2019).

ever, problematic to establish since it is extremely difficult for the AI to judge who last touched the ball if it has been deflected (Gottschalk, Tewes & Niestroj, 2020, p. 49). So, there is clear potential for the use of AI in refereeing 'black-and-white' decisions, due to the advancements in object recognition by methods of deep learning that were indicated above.

Although offside decisions are not clearly black or white, they do have an important location identification aspect to them, and this is where AI can come in: the measurability of the offside position and the moment of ball release is possible. Semi-automated offside technology was first trialled at the 2019 FIFA Club World Cup in Qatar, where limb-tracking technology was coupled with AI algorithms to determine which limb is closest to the goal line when a ball is played. FIFA further tested the use of an artificial intelligence system at the Arab Cup of 2021 in Qatar to improve the VAR impediment review process in order to make the procedure faster and more accurate. According to Johannes Holzmüller, FIFA's director of innovation, FIFA is working with companies in the artificial intelligence field to make the resource trustworthy as the delay in verifying information in the VAR generates fury in the fans, interrupts the game's progress, and generates distrust of the tool's veracity.

Currently, algorithms are being developed for predicting players in an offside position, based on an image of a specific scene in a soccer match. As the offside decision is based on visuals, computer vision techniques are a viable option for tackling these issues, by automating appropriate aspects of the process (Ushida et al 2021). One of the algorithms works according to the following steps (Panse and Mahabaleshwarkar, 2020):

1. Finding the vertical vanishing lines and vanishing point using the markings on the field to determine the relative position of all the players with respect to the field.
2. Finding the farthest horizontal projection of the playable body parts of all players (necessary for the offside rule)
3. Classifying the people on the field into Team1, Team2, and the Goalkeeper by clustering their jersey colours.
4. Finding all players in an offside position using the novel computation algorithm

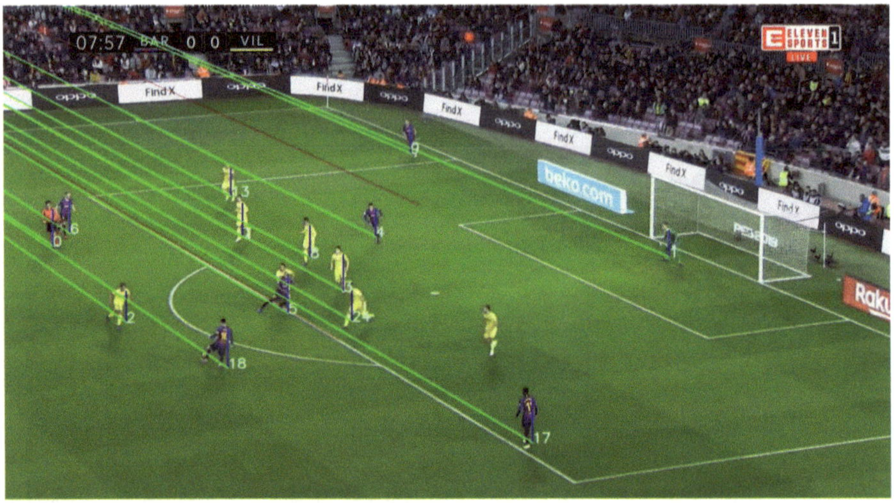

Figure 2: Example of the system working. Note: The vanishing lines (green) drawn from a single vanishing line (red). Relative positions of all players and projection of their farthest body part (vertical blue)

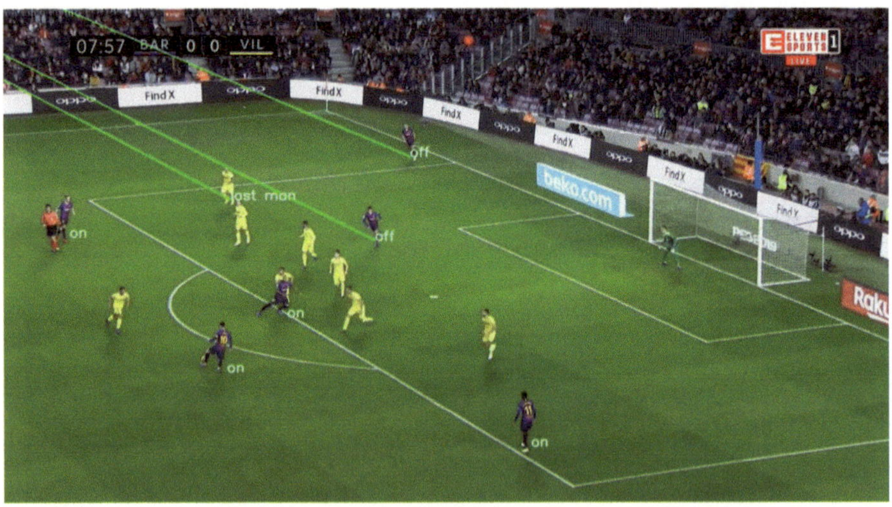

Figure 3: Example of the system working. Note: All attacking team players classified as OFF or ON based on their location. The last man of the defending team is highlighted as well.[7]

7 The presented dataset and pipeline implementation code is available at: https://github.com/Neerajj9/Computer-Vision-based-Offside-Detection-in-Soccer (accessed March 23, 2023)

AI can clearly assist decision-making on offside positions, as the offside position and the moment the ball is being played to the player in (possible) offside position (this moment is in the Laws of the Game decisive for deciding whether a player is offside or not) are easy to determine. For this powerful hardware needs to be available for trained algorithms to evaluate match scenes in real time and make AI-based decisions. However, letting the offside decision be taken fully by the AI is problematic as there are also discretionary decisions (i. e., decisions with decision-making latitude) involved that are open to interpretation. These decisions mainly concern issues of whether a player in offside position (not receiving the ball) is intervening in the game and thereby providing a teammate (that did receive the ball) a possible advantage (Gottschalk, Tewes & Niestroj, 2020, p. 49).

AI may also be useful for referee decisions on whether a defender has obviously taken away a goal-scoring opportunity for an attacker (in such a case the defender must be sent off the field with a red card), as analysis methods already work well with AI. The Expected Goals (xGoals) algorithm uses both event data and position data in a game to determine the probability of a player scoring a goal, wherein the angle of the shot, the distance of the attacker from the goal, speed, goalkeeper and defender position are compared to thousands of records in a database and subsequently can give a value (between 0 and 1) to the referee (Gottschalk, Tewes & Niestroj, 2020, p. 49).

Figure 4: xGoals as a tool for evaluating an obvious goal-scoring opportunity
Source: in Gottschalk, Tewes & Niestroj (2020, p. 38)

The last possibility for AI to assist the decision-making process in football is in case of an On-Field-Review, where AI can reduce the (oftentimes long) time the field referee needs for this. The VAR-team must decide on which camera angles are most relevant for the referee to make a correct decision, and this is a timely process, executed under high pressure, and (thus) with space for errors and interpretation. AI can help prioritise the angles on display to officials so that they only see the most relevant angles, as fewer angles mean faster decisions and/or more time to make accurate decisions.[8] By using (supervised) ML an algorithm can be built that will learn which angles are relevant and which are not. Experts, experienced referees, would go through historical footage and tag the angles that are useful and which are not, and thus provide example data and interpretations to the program so that it can learn to provide predictions.

Artificial intelligence and interpretation on the soccer field

Despite the opportunities for AI to improve the accuracy of the decision-making process, huge challenges remain. These concern first of all the acquisition of training data in the form of examples for referee decisions: for AI to learn, a vast amount of data must be available, at best from all perspectives. Also, the evaluation by AI of sequences of images is an issue, as its training is scarce so far, and the evaluation of images in order to discover parameters must still be done by humans (Gottschalk, Tewes & Niestroj, 2020, p. 48). As a result of this the speed in decision-making that is needed in an emotional, dynamic, fast-moving activity such as football is difficult to achieve. Also, as wrong referee decisions (that can be improved by AI) are oftentimes a result of incorrect perception, optimal insight into the game is required and thus more cameras above the playing field need to be implemented.

The second issue concerns decisions involving aspects of interpretation, such as on player intention and potential sabotage of the game (decisions on 'professional fouls' and yellow or red cards). These discretionary decisions concern foul play, disciplinary actions, and handball offences with grey areas and room for interpretation especially due to the expression of naturalness or the intention. Discretionary decisions are subject to a high degree of interpretation and require a sharp understanding of the situation. It shows that umpiring is not just about accurate decision-making but also about 'game management' wherein accuracy is considered less important than officiating a game without any unnecessary disruptive incidents (Bar-eli et al 2011) so that the referee can use advantage play and delayed whistle, giving a warning, and duration to manage the match. For a referee, the evaluation of the non-measurable is very important, as well as nonverbal commu-

8 https://medium.com/filament-ai/can-ai-solve-the-var-headache-8da8a7e5c286 (accessed March 23, 2023)

nication, and it is pivotal to act according to the situation to achieve the highest possible acceptance. The referee uses his empathy to assess emotions, intensity of play, and the risk of injuries, and therefore the use of AI for decisions that have to do with foul evaluation is considered difficult. Here AI reaches its limits, because the assessment of these factors is at the discretion of the referee and requires situated and experiential involvement in the game. AI can at best be deployed here to describe situations, analyse images, and draw attention to parameters that would indicate a foul play or a certain disciplinary action (Gottschalk, Tewes & Niestroj, 2020, p. 46). The challenges concern the 'situational understanding' that is required for acceptance of the decisions and a fair execution of the football activity. In order to better understand the intricate connection between the human mind and the (technological) web of relations that it is embedded in, we will now turn to Elias's study of civilisation.

2 Using Elias's psychogenetic approach

Intelligence and self-control

Artificial intelligence is often considered a mimicry of the human mind, which is considered the original substance. For Elias, however, intelligence is not some sort of substance but rather a process, a process of (increased) capacity of foresight, oversight (see Elias, 2000, p. 375), and this opens the perspective of how technologies can extend such processes. Elias's theory of civilisation rejects the idea that in Western Europe from the 16th century on, some sort of new substance, i. e., reason, is entering the stage:

> This often-noted historical rationalization is not something that arose from the fact that numerous unconnected individual people simultaneously developed from "within", as if on the basis of some pre-established harmony, a new organ ' or substance, an "understanding· or "reason" which had not existed hitherto. What changes is the way in which people are bonded to each other. This is why their behavior changes, and why their consciousness and their drive-economy and, in fact, their personality structure as a whole, change (Elias, 2000, p. 402).

Elias's theory of civilisation – that leans towards an important part of Freud's theory of human personality – is used here as a theoretical framework, which I will now describe.

Elias considers the civilising process (human development towards greater intelligence and rationality) the effect of the ever-increasing division of functions, in which it is about "the transformation of social functions and thus of conduct and the whole personality" (Elias, 2000, p. 387), that makes individuals more and more dependent on others and the (social) tissue of interdependencies. The progression of 'civilized' behaviour is according to him an effect of this tighter network of dependencies.

> This rise in the division of functions also brings more and more people, larger and larger populated areas, into dependence on one another; it requires and instils greater restraint in the individual, more exact control of his or her affects and conduct, it demands a stricter regulation of drives and – from a particular stage on – more *even* self-restraint (Elias, 2000, p. 429).

For instance, the habit of spitting, which was quite common during the Middle Ages and even a general need wherein the only limitation seemed to be not to spit over the table but under it, became slowly more distasteful in the 16th century (Elias, 2000, p. 133) – not as a result of some sort of moral development of the human spirit but due to the fact that people increasingly worked and lived in the same spaces and therefore had to change their behaviours accordingly. Civilised, intelligent behaviour is thus an increased restriction of the natural drives, and to elaborate on this crucial (psychological) theme Elias makes use of the work of Freud.

> What is decisive for a human being as he or she appears before us is neither the "id" alone, nor the "ego" or "super-ego" alone, but always the relationship between these various sets of psychological functions, partly conflicting and partly co-operating levels in self-steering. It is these relationships within individual people between the drives and affects that are controlled and the socially instilled agencies that control them, whose structure changes in the course of a civilizing process, in accordance with the changing structure of the relationships between individual human beings, in society at large (Elias, 2000, p. 409).

According to Elias, following Freud, drive energies are controlled by moral expectations (the super-ego), and thus formed and moulded by them. But for him these libidinal energies (the *Es*) are not some sort of timeless and universal energetic reservoir, as Freud thought, but different in different historical epochs wherein the tissue of interdependencies of people is different.[9]

---

9 Elias critiques Freud for this: "No distinction is made between the natural raw material of drives, which indeed perhaps changes little throughout the whole history of humankind, and the increasingly more firmly wrought structures of control, and thus the paths into which the elementary energies are channelled in each person through his or her relations with other people from birth onward" (Elias, 2000, p. 409).

> Therefore, in order to understand and explain civilizing processes one needs to investigate – as has been attempted here – the transformation of both the personality structure and the entire social structure. This task demands, within smaller radius, psychogenetic investigations aimed at grasping the whole field of individual psychological energies, the structure and form of the more drive-impulsive no less than of the more conscious self-steering functions (Elias, 2000, p. 411).

In this psychogenetic approach individual behaviour is only understandable as an effect of a certain phase, "as a part of a particular stage or wave" (Elias, 2000, p. 403), and is the result of the (social) tissue of interdependencies. Changes in human behaviour are not a result of increased intelligence or insights but of a different structuration of the drives as a result of tighter-knit social interaction.

Pivotal is Elias's diagnosis that the progression towards more civilised behaviour in human history *implies at the same time more self-control* of the individual.

> The basic tissue resulting from many single plans and actions of people can give rise to changes and patterns that no individual has planned or created. From this interdependence of people arises an order sui generis, an order more compelling and stronger than the will and reason of individual people composing it (Elias, 2000, p. 366) [...] Individuals are compelled to regulate their conduct in an increasingly differentiated, more even and more stable manner (Elias, 2000, p. 367)

This distancing (differentiation) inside the personality, between the 'natural passions' and the goals and ideas of a more deliberate (conscious, calculative, reflected) attitude (appearance), is actually, according to Elias, the (psychogenetic) core of the civilising process:

> what we refer to by the reifying terms "reason", "ratio" or "understanding" (...) these terms express a particular moulding of the whole psychic economy [...] They are aspects of that moulding by which the libidinal centre and the ego-centre are more and more sharply differentiated, until finally a comprehensive, stable and highly differentiated agency of self-constraint is formed (Elias, 2000, p. 402–3).

This shows that the work of reason, which is one of increased foresight and control, always leaves its 'wounds' in the person, and the price to be paid for increased civilised behaviour and interaction is increased self-control. "The learning of self-controls, call them 'reason' ... or 'conscience', 'ego' or 'superego', and the consequent curbing of more animalic impulses and affects [...] is never a process entirely without pain; it always leaves scars" (Elias, 2000, p. 377). For "self-restraint (...) is the price, if we may call it so, which we pay for our greater security and related advantages" (Elias, 2000, p. 429).

Elias's framework used for understanding the effects of AI

By analysing AI in football from Elias's perspective the chapter tries to disclose how the (individual) human being is altered at the level of the passions and the 'pleasure balance' (Elias, 2000, p. 378). This approach is congruous to a line of critique on the VAR stating that it makes the game 'more passionless.' Elias helps to see how making the game fairer, more civilised, so to speak, is at the same time making it more passionless: the use of AI-led VAR (the term used from now on to refer to refereeing assisted not only by video but also by AI) for increased 'fairness' in football converges with increased 'self-control'. The engine of this is the order of (technological) interdependencies that no individual has control over.[10]

In AI, intelligence emerges when an intelligent agent or computer system perceives its environment in order to (take actions to) achieve certain goals. This intelligence is at work on the football pitch when the umpiring team is expanding its natural perception of the environment with technologies of computer vision in order to (better) achieve its goals of accurate decision-making. In the psychogenetic approach these improved processes for perceiving and decision-making are not only operating at the level of neutral information processing (better insights and decisions due to better information processing capabilities) but also have direct effects at the affective level of passions and drives (and with that to the core of the human personality).[11] Technologically enhanced decision-making on the pitch (unconsciously) subjects referees more to mechanisms of (self) control – so that, for instance, the quirky, idiosyncratic referee becomes an anachronism.

The psychoanalytic stance on progress is one of being also sensitive to the shadow sides of this advance of civilisation (that is why Freud wrote his *Civilization and its Discontents*). One of the laws discovered in psychoanalysis is that increased control of someone's affective life (by means of the conscience of super-ego structures) leads to increased internal tension. Elias uses Freud to diagnose that "these constraints also produce peculiar tensions and disturbances in the conduct and drive economy of the individual. In some they lead to perpetual restlessness and dissatisfaction, precisely because the person affected can only gratify a part of his or her inclinations and impulses in modified form, for example in fantasy, in looking-on and overhearing, in daydreams or dreams" (Elias, 2000, p. 376).

---

10 Elias refers to Hegel's concept of the 'cunning of reason' to illuminate that from all the interdependencies between human agents, their planning and actions, may emerge an order of things that no one actually intended (Elias, 2000, p, 366).

11 "But any investigation that considers only people's consciousness, their reason or ideas, while disregarding the structure of drives, the direction and form of human affects and passions, can from the outset be of only limited value" (Elias, 2000, p. 408).

The aim here is to study the effects of AI on human behaviour and human self-experience that are probably the hardest to recognise, as those effects are so intimately bound to 'normal' ways of experiencing ourselves and the world that they tend to be overlooked. One easily gets used to the possibilities and effects of technologies, and therefore I now try to open up this adapted (modified) behaviour – by looking at its magnified manifestations in the world of football.

3 The psychic effects of artificial intelligence in the football game

AI and changed forms of enjoyment: the rationalisation of the spectator experience

AI in sports is used for talent identification and selection (processing data on players' performance to predict their potential and market value); pre-game preparation (such as processing data on nutrition, biomechanics, mental and physical aspects of players, all used for training, coaching, strategy, injury management, and team selection); and post-game analysis (Di Stefano, 2021). AI also increases media and fan experiences and marketing strategies. By analysing fan reactions and visual data AI can choose the most exciting situations to be broadcast on television and the Internet; it can revolutionise the way sports storytelling is done by using machine learning and deep learning algorithms to automate videomaking operations; it can offer fans additional stats and insights to enrich their experience and ensure better customer support (Di Stefano, March 23, 2021). AI can also improve marketing effectiveness by offering advanced targeting based on fan demographics, including media consumption behaviours, personal interests, and shopping habits.

The sports system is increasingly penetrated by intelligent technologies, so its 'elements' of players, referees, coaches, spectators, media channels, laws (of the game), commercial interests etc. are more and more influenced by them. It shows the increased interconnectedness of all the elements involved, or what Elias describes as "the interdependence of larger groups of people" (Elias, 2000, p. 375). The experience of the spectator shifts more to the 'reflective' experience of reviews, data, and steered marketing and customer experiences. So, in all these developments the libidinal economy of the spectator changes (or needs to change) as well. This is best visible in the changed position of the spectator after the scoring of a goal: there needs to be (increased) distancing towards this 'immediate reaction' or impulse (a cheerful reaction of joy or one of dismay when the opponents have scored) as there is a 'secondary time' of VAR evaluation of the goal in order to make sure that no irregularity or accident took place in the original event, and spectators must 'delay' their emotional reaction to after this secondary evaluation, must hold and keep down their joy and emotions as they need to await whether the goal will really be approved after the checks. Only in a 'secondary time' the goal is 'objectively' awarded: a joy or (drive) fulfilment according to the (new, technology-led) codes. This delay exemplifies the (further) delay of drive fulfilments that Elias assigns to increased civilised behaviour, wherein these "self-constraints, [are] a function of the perpetual hindsight and foresights" (Elias, 2000, p. 375).

A different drive-economy – or, in the words of French psychoanalyst Jacques Lacan, a different kind of 'enjoyment' – arises: a drive-economy with a larger focus on data-led reviewing instead of immediate (in-game) experience. A football fan must accept this new economy of emotions as the condition (or price to be paid) for increased fairness, 'objectivity', measurability, and overview in the game – it shows how people's drive-economy and their personality structure depends on "the way in which people are bonded to each other" (Elias, 2000, p. 402). And he may develop new forms of enjoyment based on this changed regime: enjoying the reviews, the game statistics and data, abilities to purchase merchandise, or communicating on social media on game-situations.

AI and the self-distancing of more intelligent umpiring

Along with the rising pressure exercised by (television, internet) audiences due to increased opportunities for reviewing a situation, the demand for more or fully accurate decisions grows. Viewers at home see each move and referee error in replays, and this mounts the pressure for more accuracy in the decision-making process. Television viewers were oftentimes, and especially due to reviews and replays, in a better position than the referee to evaluate and oversee actions and situations, which led to a growing sense of unease and distrust among viewers, so in order to change this growing distrust, decision-making in sports had to become more accurate to regain trust in the spectator experience.[12] Where traditionally the decision-making process on game-activities was supposed to take place best on the field, with the growing role of spectator technologies the rules changed.[13]

12 Tamir & Bar-eli, 2021, p. 5.
13 See Collins (2019).

Also due to the huge financial interests involved in professional football errors in umpiring are considered 'no longer tolerable'.

AI-driven VAR is directed by the goal of increased 'rationality' in the abidance to and implementation of rules during the football match: increased fairness. Practically this means that after having taken a decision in the dynamic context on the pitch, the referee may be called back by the VAR-team in order to take another look at the situation, and possibly reconsider his earlier decision (which was based on 'insufficient' information and lack of oversight due to the immersion in the dynamic context on the pitch). The decision-making process is thus brought on a 'higher level' of a more distanced, better informed, and more abstract view. This whole process of technology-enhanced decision-making demands the referee (team) to take a larger distance towards their 'natural inclinations' (decision-making based on more direct impressions of the situation at hand) in order to improve the fairness of the game.

To put it in Elias's terms, these technologies advance 'civilisation' in (professional) playful activities by diminishing the 'violence' of injustice: detecting game violations. With that, they bring the playful activity more under the rule of abstraction, as the implementation of the rules of the game becomes less dependent on situational understanding and on local circumstances and culture.[14] Local differences should not influence the decision-making process as the rules of the game should be implemented everywhere the same (where the problem of erasing these differences becomes especially visible in European matches where teams of different countries engage): the use of intelligent technologies is the guarantor of such an 'abstract', more neutral implementation of the rules.

This striving for increased abstract and neutral decision-making is also noticeable in the infield decision-making process of referees, as more and more referees seek to become very strict and 'clear' in their implementation of rules: for instance, giving red cards for minor violations that are under the exact rules of the game and especially after repetitive review (i. e. abstract), considered major offences. Or not only giving a penalty for a handball in the penalty area but also sending the player off even when the handball was not intentional and when it is early on in the game so that sending off the player has a major impact on the further game, often disturbing it to a large degree. However, such a 'coincidental' or 'timely' circumstance should not play a role in the 'abstract' application of rules. Being firm in the application of abstract rules is desirable as activities can be re-

14 The English Premier League is renowned as the most physical one in the European association football, whereas, in contrast, the French Ligue 1 is less intense (SkillCorner, 2020). What is interesting is that critique on VAR technologies is most intense in England.

viewed, also by referees' superiors in the umpiring boards who demand exact application of the rules so that a referee, in order to advance his career (which shows how an obviously unambiguous activity imbibes an intricate network of interdependencies), must show uttermost compliance and willingness to be strict.

The goal is diminishing elements of subjectivity and bias, defined as 'distortion of measurement' or 'evaluating results leading to their misinterpretation' (Helsen, MacMahon, and Spitz 2019) and to bring the rationality of the decision-making process on the pitch to an objective, neutral point of 'correct interpretation'. It shows how the use of AI is (implicitly) driven by the (Enlightenment, rational) ideal of a neutral, distanced point of view from which reality (in this case on the football pitch) can be viewed 'as it really is'. From this Enlightenment perspective of increased civilisation and rationalisation, all the various (local, situated) factors influencing the decisions of arbitrators should be eliminated as far as possible.[15] Further examples of these factors influencing arbitrators are the home crowd's noise (Nevill, Balmer, and Williams 2002), players' reputations or a team's origin (Gottschalk Tewes & Niestroj, 2020), or the referees' own prior decision (Brand, Plessner, and Schweizer 2009). Such influencing factors can lead to what is called a compensation bias, so referees who make decisions in favour of one team in subsequent decisions try to even out the situation (Tamir and Bar-eli 2021, p. 3). Of course, also the angle of a referee's sight of the ball, of the players, or of assistant referees can lead to biased decisions (that's why the use of AI to help the referee team in providing the most significant camera angles was discussed). All such "influencing factors, as well as subconscious sympathies or antipathies, would be neutralized by using AI for the referee" (Gottschalk Tewes & Niestroj, 2020 p. 45).

The aim (dream of reason) penetrating AI is to represent 'true reality'. In the case of the football pitch, it means that what is *really* going on, i. e., the *correct* interpretation of the situation, is as it is observed by means of intelligent technologies. The introduction of new technologies, here in a playful activity on a football pitch, dictates new standards for evaluating reality (McLuhan 1964; Tamir and Bar-eli 2021, p. 6). Progress (in intelligent technologies) assumes that we (as engaged spectators, players, referees) are all lured by our bodily and emotional perception of the situation and *should* – this is the demand of reason, progress, fairness, and civilisation – subject this immediate perception to the calculation offered by the visual technologies and smart algorithms. Actually, at an ontological level (i. e., reflecting on what is reality and what is illusion) there is an (implicit) claim that the

15 Due to its demand of eliminating 'differences' this rationalistic progress is penetrated by forms of violence.

evaluations on the pitch, i. e., situational and taking place in time, are inferior to the interpretation from a point of view beyond time and spatiality. The ultimate goal of the use of AI in football is to integrate these two levels (of situated and disembodied rationality) in order to overcome the major issue of interrupting the flow of the game so that the football activity can move on without restraint while at the same time being subjected to the sublime evaluation and interpretation offered by intelligent technologies – the referees' point of view on the pitch must be combined/connected to the distanced view of the AI-driven VAR. Then 'body' and 'mind' would be reunited again on a higher (moral) level: the ultimate dream.[16]

Intelligent player behaviour

The psychodynamic analysis brings to light an amounting pressure on actors. Financial pressure: football players being particularly prone to illegal behaviour because they are under enormous pressure of the game and of losing their multi-million-dollar salaries (Aarnink, 2021, p. 4). And technological pressure: due to a high-processing capacity for reviewing, correctly abiding by the rules becomes more prevalent, and this stricter implementation of the (self-)control apparatus changes the behaviour of those involved. Also, as a result of the increased visibility of violations, players are increasingly posing and shamming in their behaviour in order to influence the referee team and be awarded a free kick for an offence. As minor details of an action are noticeable and stand out, a minor offence already tends to be evaluated (from the elevated perspective that we increasingly identify with) as a violation of the rules, and players seek to take advantage of this changed rule implementation. The stricter the VAR is in penalising physical intensity, the more players are encouraged to show sabotage behaviour, actions that exert a negative impact on rival performance such as seducing the referee into awarding a penalty or free kick or giving the opponent a yellow or red card (Aarnink, 2021, p. 22). The (interiorised) intelligence of players, i. e., their capacity to have more insight and foresight in what is useful to them in achieving success in the game (their rationality), is changing as they are learning about what kind of behaviour is bringing results.

16 This dream bounces into the 'hard rock' of the real whether some situations, for instance, on player intention and potential sabotage of the game such as decisions on 'professional fouls' and yellow or red cards, might not best be evaluated in the flow and full context of the game. However, this implies recognising the limitations of the dominant regime of calculation.

Repetitive reviewing changes the meaning of the action: meaning shifts from direct action on the pitch towards what it means for the distanced point of view of the VAR-team, i.e. the indirect, moral perspective of 'objective' rule implementation (where Freud already taught that the moral domain is exactly this distanced evaluation).[17] As such, AI-driven VAR technologies represent a further 'virtualization of action' – as the implementation of rules is always a virtualization of action where the 'violence' of direct action is brought under the rule of law (Levy 1998). So just the bare fact of using more intelligent technologies for umpiring already has the effect of bringing the (playful) activities more under a moral perspective.

As the game can be reviewed strictly (scrupulously, accurately, conscientiously), the (self-)control apparatus is gaining more influence on players: they need to restrain themselves in the intensity of their actions so that not only their behaviour changes but even their personality(structure). The earlier 'warrior' type of football player (such as in the heydays of Dutch football in the 1970s players like Johan Neeskens and Willem van Hanegem) finds a hard time to survive in technologically surveyed play nowadays. The current player needs to be 'smarter' (instead of showing old-fashioned physical contact) and may show intelligent behaviour to such an extent that it ends up doing all that is needed – exaggerated falling, screaming at the slightest touch – for being awarded a foul. That type of player is stimulated that can gain the "advantage of those able to moderate their affects" (Elias, 2000, p. 370). It fits in with Elias's diagnoses that the personality structure changes and "the modelling of the individual self-steering apparatus" [...] occur "wherever functions are established that demand constant hindsight and foresight in interpreting the actions and intentions of others" (Elias, 2000, p. 378–379). AI-driven VAR technologies require that "individuals are compelled to regulate their conduct in an increasingly differentiated, more even and more stable manner" (Elias, 2000, p. 367). Also, Tamir and Bar-eli conclude in their study that the

> VAR system, which receives (unsurprisingly) much resistance, should be treated as one of the most important and moral changes in soccer [...] that requires an internalization by human beings to comprehend wrongfulness, wilfully alter their attitudes and then express right actions (Tamir and Bar-eli, 2021, p. 6)

---

17 The question concerning human behaviour, from a psychoanalytic perspective, is what is directly 'natural' or even 'instinctual' about it (let's call this the Freudian *Es*) and what is adapted or inhibited 'social' or 'moral' about it (the Freudian super-ego, or what Elias calls the apparatus of self-control) – and what is a healthy relation between the two (where Freud, for instance, in his time diagnosed 'Victorian morality' as unhealthy).

The drive-economy of the players changes as they are more restrained in unleashing their 'aggressive' energies on the pitch, and one of the general fears on the effects of (AI-driven) VAR technologies is that football becomes a less passionate sport, not only because players and fans can't celebrate goals properly, but also because of this inhibiting factor. Life, in this case life on the pitch and on the screen (of spectators), "becomes in a sense less dangerous, but also less emotional or pleasurable, at least as far as the direct release of pleasure is concerned" (Elias, 2000, p. 375). This shows a remarkable resemblance to the progress of civilisation that Elias describes and wherein the direct confrontation of the warrior type needs to be sublimated due to a stricter moral regime of interdependencies ('The Courtization of the Warriors' is part 4 of Elias's Synopsis of the Civilising Process). Elias also ascertains that these inhibited energies subsequently seek new avenues and outlets (remedies), for instance, in books (wherein the heroic actions are now symbolically made present), as for what is lacking "a substitute is created in dreams, in books and pictures" (Elias, 2000, p. 375). Similar exits may be diagnosed for the professional football players (and spectators) that seek to find exits for their impulses on social media and satisfy their impulses by finding stardom and fame there: many professional football players are celebrities on social media.

So, the advanced use of AI-driven VAR technologies has the objective of making the game fairer but at the same time threatens to make it 'passionless'. This brings us back (see note 16) to the question concerning the 'healthy' relation between affects and their regulation, and guiding herein may be Elias's conclusion of his civilisation theory "that the common pattern of self-control expected of people can be confined to those restraints which are necessary in order that they can live with each other and with themselves with a high chance of enjoyment" (Elias, 2000, p. 446). This brings us to the question regarding (post)human existence

4 Human, transhuman, posthuman?

The goal and future of AI-driven VAR technologies is to ensure increased rationality and fairness in the decision-making process. I already diagnosed this striving (will, desire) as driven by the Enlightenment ideal of observing and evaluating reality from a 'neutral', distanced point of view – for this is the ideal when a referee leaves the 'situated action' at the pitch for observing what 'really was the case' in the VAR box. Thus, enhancing the decision-making process in football is part of the transhumanist movement to 'improve' and enhance the human, and to transcend the limitations of the human by using technologies. Relating to athletes' performance, for instance, this enhancement is studied quite extensively under concepts such as performance enhancement, techno sport, human enhancement technologies, or mechanical ergonomics. When connected to issues of unfairness or negativity surrounding its use it is called technological doping or techno doping (Dyer 2015). The enhancement of the decision-making process is transhumanist because human rationality is considered the key marker of personhood, so the intention to improve human activities is a matter of enhancing this rationality (Nayar 2014, p. 5–10).[18] The VAR intends to create a more rational and 'morally advanced' form of human play.

The attempts of improving human activity are governed by the ideal of a better and more civilised reality – both a more elevated physical world and a more sublime time. Concerning the first, physical dimension the referee's situated, bodily, and thereby limited perception is the problem that should be overcome by enhancing perception with AI-driven VAR technologies. These should improve his rationality, authority, autonomy, and agency: all the characteristics that Western humanist rational tradition attributes as essential to the human (the transhumanist is therefore the technological continuation of this tradition). Also, the players should adapt their physical, bodily inclinations and subject themselves to this new technological regime of decision-making.

As to the second dimension, time, the ideal is that of integrating intelligent decision-making technologies thus into the game so that the annoying delay and disruption is eliminated: rationally enhancing the decision-making processes while at the same time maintaining the 'natural' flow of the game. Then the gap ('delay') between the primary time of the activity and the secondary time of its evaluation is closed, and the technical becomes a 'natural part' of the natural world – *without this integration standing out clearly.* That would bring all the participants of the game in a posthuman condition, wherein the technical is no longer an enhancement or prosthesis but an *integral part* of it. In this posthuman condition 'body' and 'mind' would be reunited again on a higher (moral) level: its ultimate dream.

Thus, the play of football shows itself to be a *co-evolving process* of the human and the technical, wherein all the elements (players, referees, spectators, etc.) adapt and evolve their ideas and mindsets to technologically mediated environments that are (increasingly) perceived and understood as natural. This process is – as Elias analysed – not steered and imposed somewhere from above (where at first sight it seems that a governing body like FIFA is doing this) but is slowly

18 In the discussions on transhumanism its (all too) abstract, distanced position is often addressed: jumping over the proper dynamics, energies, and differences of the world. Furthermore, the question is which cultural environment is underpinning the so-called neutral and objective point of view (is Western rationality privileged; is this rationality universal?). Applied to football: which refereeing style (English, French, etc.) should govern the VAR technologies, or is there an 'objective and universal' style?

evolving because of the changing interdependent relations. Technological developments are changing the relations between human individuals, adapting their rationality and their drive-economies. Unconsciously, ideas and expectations, experiences of self and body are transformed. A notion like 'masculine game' and the idealisation of 'rough' players such as Neeskens and Van Hanegem will become something of a previous, less civilised era. The technologically (rationally) mediated experience and evaluation of reality has become its 'natural' interpretation. As such, AI-driven technologies in football are a next step in the civilising process.

However, the question is whether a troubling insistence on a 'human' or 'natural' world (that evades this process of techno-rationality) does not remain. In football, discussions flare on whether the interpretation of some situations might not best be made in the flow and full context of the game; on possible misinterpretations due to 'distanced' video replays (however fast they may be with AI); on compensating (bias) as something that is not necessarily a deviation to be eliminated but sometimes necessary for a fair game; on the value of different refereeing styles; on possible dehumanisation of football wherein human mistakes are a facet of both sport and everyday life; and on refereeing as navigating also on instincts rather than merely the outright objectivity of the facts (Svantesson 2014). To conclude philosophically: thinking that all these 'differences' can be brought under the rule of the (moral) universal is a (Hegelian) dream.

Literature

Aarnink, A. (2021). How Does the Video Assistant Referee (VAR) Affect Players' Sabotage Behavior?. Available at SSRN 3889340.

Bar-Eli, M., Plessner, H., and Raab, M. (2011). *Judgement, Decision Making and Success in Sport.* Chichester: Wiley-Blackwell.

Brand, R., Schweizer, G., & Plessner, H. (2009). Conceptual considerations about the development of a decision-making training method for expert soccer referees. *Perspectives on cognition and action in sport*, 181–190.

Collins, H. (2019). Applying philosophy to refereeing and umpiring technology. *Philosophies* 4, 21–27. doi: 10.3390/philosophies4020021

Di Stefano, A. (March 23, 2021). AI in sports: current trends and future challenges. https://www.itransition.com/blog/ai-in-sports

Dyer, B. (2015). The controversy of sports technology: a systematic review. SpringerPlus, 4(1), 1–12.

Elias, N. (2000). *The civilising process. Sociogenetic and psychogenetic investigations.* Malden, Ma./Oxford: Blackwell Publishing.

Gelberg, N. (1998). Tradition, talent, and technology: the ambiguous relationship between sports and innovation. *Sports in design*, 88–110.

Gottschalk, C., Tewes, S., & Niestroj, B. (2020). The innovation of refereeing in football Through AI. *International Journal of Innovation and Economic Development*, 6(2), 35–54.

Helsen, W. F., MacMahon, C., & Spitz, J. (2019). Decision making in match officials and judges. *Anticipation and decision making in sport*, 250–266.

Huang, W. (2020). Discussion and Analysis on the Application of Artificial Intelligence in the Field of Competitive Sports. In *Innovative Computing* (pp. 737–742). Springer, Singapore. https://doi.org/10.1007/978-981-15-5959-4_90

Jones, M. V., Paull, G. C., & Erskine, J. (2002). The impact of a team's aggressive reputation on the decisions of association football referees. *Journal of sports sciences*, 20(12), 991–1000.

Lévy, P. (1998). *Becoming Virtual. Reality in a Digital Age.* New York / London: Plenum Trade.

Loland, S. (2002). Technology in sport: Three ideal-typical views and their implications. *European Journal of Sport Science*, 2(1), 1–11.

McLuhan, M. (1964). *Understanding Media: The Extensions of Man.* New-York, NY: McGraw Hill.

Nayar, P. (2014). *Posthumanism.* Cambridge: Polity Press.

Nevill, A. M., Balmer, N. J., & Williams, A. M. (2002). The influence of crowd noise and experience upon refereeing decisions in football. *Psychology of sport and exercise*, 3(4), 261–272.

Panse, N., & Mahabaleshwarkar, A. (2020, October). A Dataset & Methodology for Computer Vision based Offside Detection in Soccer. In *Proceedings of the 3rd International Workshop on Multimedia Content Analysis in Sports* (pp. 19–26). https://doi.org/10.1145/3422844.3423055

Pretorius, A., & Parry, D. A. (2016, September). Human decision making and artificial intelligence: a comparison in the domain of sports prediction. In *Proceedings of the Annual Conference of the South African Institute of Computer Scientists and Information Technologists* (pp. 1–10). https://doi.org/10.1145/2987491.2987493

Raab, M., Bar-Eli, M., Plessner, H., & Araújo, D. (2019). The past, present and future of research on judgment and decision making in sport. *Psychology of Sport and Exercise*, 42, 25–32.

Rathi, K., Somani, P., Koul, A. V., & Manu, K. S. (2020). Applications of artificial intelligence in the game of football: The global perspective. *Researchers World*, 11(2), 18–29.

Sabag, E., Lidor, R., Morgulev, E., Arnon, M., Azar, O., & Bar-Eli, M. (2020). To dive or not to dive in the penalty area? The questionable art of deception in soccer. *International Journal of Sport and Exercise Psychology*, 18(3), 296–307.

SkillCorner (2020). Let's Get Physical: Comparing the Big 5 European Football Leagues. https://medium.com/skillcorner/lets-get-physical-comparing-the-big-5-european-football-leagues-bcb04ebb835c

Spitz, J., Wagemans, J., Memmert, D., Williams, A. M., & Helsen, W. F. (2021). Video assistant referees (VAR): The impact of technology on decision making in association football referees. *Journal of Sports Sciences*, 39(2), 147–153.

Svantesson, D. J. B. (2014). Could technology resurrect the dignity of the FIFA World Cup refereeing?. *Computer Law & Security Review*, 30(5), 569–573.

Tamir, I., & Bar-Eli, M. (2021). The moral gatekeeper: Soccer and technology, the case of Video Assistant Referee (VAR). *Frontiers in psychology*, 11, 613469.

Uchida, I., Scott, A., Shishido, H., & Kameda, Y. (2021, October). Automated Offside Detection by Spatio-Temporal Analysis of Football Videos. In *Proceedings of the 4th International Workshop on Multimedia Content Analysis in Sports* (pp. 17–24). https://doi.org/10.1145/3475722.3482796

Zhang, J., & Li, D. (2021, August). The Application of Artificial Intelligence Technology in Sports Competition. In *Journal of Physics: Conference Series* (Vol. 1992, No. 4, p. 042006). IOP Publishing.

Roberto Redaelli

From Tool to Mediator. A Postphenomenological Approach to Artificial Intelligence

Abstract: This chapter aims to clarify the moral status of AI systems by applying to them the notion of moral mediator developed by P. P. Verbeek in the field of Science and Technology Studies (STS). More precisely, we propose to define artificial intelligent systems as moral mediators of a particular kind, i. e. as possessing technological intentionality linked to composite intentionality. To this end, it is first necessary to show that the common view of technology held by various forms of instrumental theories is insufficient for the purpose of understanding the agency of AI systems. We then analyse some paradigmatic positions that assign a certain moral status to technological artefacts, such as those of D. G. Johnson and J. Sullins, in order to compare them with Verbeek's postphenomenological approach. Finally, we illustrate how this latter approach overcomes certain limitations that can be ascribed to these other positions and offers a contribution to the process of understanding the moral significance of artificial intelligence.

Keywords: postphenomenology; Artificial Intelligence; ethics of technology; science and technology studies

The problem of the moral status of artificial intelligence

The spread of artificial intelligence is radically redefining our relationship with technology: from assistive robotics to computerised decision support tools widely used in the medical and legal fields to vehicles with different levels of autonomy, these devices play a central role in both our private and professional lives. Indeed, these AI systems take care of us, guide us in our choices, and facilitate the performance of the many tasks we are called upon to perform in our daily lives.

Given the services offered by these machines, increasingly urgent questions are raised today about the ethical implications of the pervasive use of such systems in our society: can robots replace humans in caring for the elderly or educating

https://doi.org/10.1515/9783111007564-006

young children[1]? What tasks can they perform? If decisions made on the basis of the guidance offered by so-called decision support systems are effective, for example, in the field of business, can we say at the same time that those decisions are ethically sustainable[2]? And again: if the widespread use of self-driving cars will lead to a reduction in the number of road accidents in the near future, who will be responsible for damage caused by these vehicles to people or property[3]?

These questions, which affect different areas of our lives, are at the centre of an ethical reflection that extends its field of enquiry to technical artefacts in the present day, and thus its focus is not only on the human and non-human living being but also on the non-living being. By examining technical artefacts, in fact, ethics redefines its boundaries, thus overcoming the various forms of ostracism that excluded, and sometimes still exclude today, both animals and machines from the community of moral subjects[4].

This new direction inaugurated in the ethical field can be witnessed in the recent debate on the moral status of artificial intelligent systems (see Coeckelbergh 2020a, pp. 47–62). Depending on the different perspectives taken, ranging from a functionalist approach (Wallach and Allen 2009) to a relational approach (Coeckelbergh 2014), different types of moral standing are ascribed to such systems: for some authors, intelligent systems are moral entities, but not moral agents because they lack mental states and intentions to act (see, e. g., Johnson 2006); for others, they are moral agents by virtue of the so-called mindless morality (Floridi and Sanders 2004); for another group, intelligent systems are mere tools which, although they do not possess any special moral status, have significant ethical effects on our lifestyles (see, e. g., Peterson and Spahn 2011). It is in this nuanced debate

---

1 The use of robots in elderly care raises a number of ethical issues that have been addressed, for example, in Sparrow 2016; Sparrow and Sparrow 2006. On the role artificial intelligence can play in educational processes, see the text *BEIJING CONSENSUS on artificial intelligence and education*, published by the United Nations Educational, Scientific and Cultural Organization (UNESCO) in 2019.

2 For a mapping of ethical issues arising in the business field, see Daza and Ilozumba (2022).

3 On this issue, see, e. g., Loh and Loh (2017); Millar (2017).

4 A common justification for the exclusion of animals and machines from the domain of morality is traditionally offered by the Cartesian doctrine that recognises them as entities that behave mechanistically. In this context, Johnson observes that "the Cartesian idea is that animals, machines, and natural events are determined by natural forces; their behavior is the result of necessity. Causal explanations of the behavior of mechanistic entities and events are given in terms of laws of nature. Consequently, neither animals nor machines have the freedom or intentionality that would make them morally responsible or appropriate subjects of moral appraisal. Neither the behavior of nature nor the behavior of machines is amenable to reason explanations and moral agency is not possible when a reason–explanation is not possible" (Johnson 2006, p. 199).

that the present contribution takes its place, with its aim to highlight the moral significance of artificial intelligence systems, applying to them the notion of moral mediator developed by P. P. Verbeek in the field of *Science and Technology Studies*. In this sense, we propose that AI agents be defined in terms of moral mediators of a particular kind, i. e. equipped with a technological intentionality linked to composite intentionality.

To this end, it is first necessary to make clear how the common view of technology held by various forms of instrumental theories (see, e. g., Pitt 2014) is insufficient for the purpose of understanding the agency of AI systems. We will then analyse some paradigmatic positions that assign a certain moral status to technological artefacts, such as those of D. G. Johnson (2006) and J. Sullins (2006), in order to show how Verbeek's postphenomenological approach overcomes certain limitations that can be ascribed to these positions and offers a contribution to the process of understanding the moral significance of intelligent systems.

Guns don't kill people, people kill people with guns: The instrumental theory and artificial intelligence

The common view of technology, the one that informs our daily lives, recognises the variety of technological artefacts as mere tools at our service. From smartphones to cars, from the hammer to the refrigerator, technical artefacts are, for us, means to our ends. In this sense, these tools are morally *neutral*, i. e. they cannot be assigned a value, but it is the use we make of them that can be subjected to moral evaluation.

A staunch defence of the so-called *value neutrality thesis* (VNT) is offered today by Joseph Pitt, for whom "technological artifacts do not have, have embedded in them, or contain values" (Pitt 2014, p. 90). In order to support this thesis, according to which artefacts do not incorporate values, Pitt cites, among various examples, the famous slogan of the National Rifle Association of America: "Guns don't kill people, people kill" (see Pitt 2014, p. 89). Behind this slogan, one can clearly recognise an application of the value neutrality thesis: it is the *use* of the firearm and not the firearm itself that has a moral value. In fact, the firearm does not exert any coercive force that drives the individual to commit a crime, but it is the holder of the firearm who decides whether and how to use it. Thus, although Pitt admits that a multitude of value-laden decisions are involved in the process of designing, producing, and marketing a technical artefact (Pitt 2014, p. 97), this does not mean that such artefacts are themselves value laden. With this thesis, he can conclude that “guns don’t kill, people kill using guns, knives, their hands, garrottes, automobiles, fighter planes, poison, voodoo dolls etc. The culprit is people” (Pitt 2014, p. 102).

Another argument employed by Pitt in order to affirm the impossibility for technologies to incorporate moral values calls into question what we shall refer to as the *location criterion.* Opposing Langdon Winner’s (1986) famous analysis of the Long Island overpasses designed by architect Robert Moses – which Winner claims embody power relations and political properties – Pitt asks, looking at such overpasses, where the values they embody are located[5]. In this sense, he asks where such values can be localised in the artefact, only to conclude that in no way do such values belong to the artefact itself since no localisation of the axiological dimension within their material conformation is possible. Such values belong to Moses and cannot be incorporated by the artefact.

Now, a number of objections have been raised to the thesis of the moral neutrality of technology, among which some particularly useful for our purposes were those advanced by D. Ihde’s postphenomenological perspective and later taken up by Verbeek. One argument in particular seems to call into question what this thesis support and what Pitt reaffirmed. In a famous passage in *Technology and the Lifeworld* (1990), Ihde points out, in an almost aphoristic tone, that “the human-gun relation transforms the situation from any similar situation of a human without a gun” (Ihde 1990, p. 27).

With this statement, the philosopher first of all criticises a fundamental assumption of the thesis regarding the supposed neutrality of technologies, namely the idea that technologies are things-in-themselves, isolated objects (Ihde 1990, p. 26). Technologies are not mere objects, but rather make up that “primitive unit” (Ihde 1990, p. 27) that is created by the human-technology relationship, and we would add that it is within this relationship that it is possible to understand the moral significance of technologies. Therefore, returning to the NRA slogan, the possession of a firearm does not reduce the weapon itself to a mere means at our disposal, but rather, for Ihde, the weapon modifies, *mediates,* our relationship to the world and thereby also our perception of ourselves. In this sense, a man

5 Specifically, Pitt states: “Let us say we have a schematic of an overpass in front of us. Please point to the place where we see the value. If you point to the double-headed arrow with the height of the overpasses written in, you have pointed to a number signifying a distance from the highway to the bottom of the underpass. If you tell me is Robert Moses’ value, I will be most confused”. Pitt continues: “if we can’t locate them [the values], then is merely *metaphorical* to say of an object that it embodies human values”. Therefore, “*if* Robert Moses had certain prejudices [...] it is Moses’ intentions, desires, i. e., values, that put certain structures in place, but that does not mean that the structures themselves have his values” (Pitt 2014, p. 94–95).

who possesses a firearm has a different relationship with the world than a man who does not possess a weapon. In the eyes of the former, we might say, other entities become potential targets on which he can rage without having to have any direct contact with them. In this way, the firearm, like any other technical artefact, reduces and amplifies certain aspects of reality, offering man different possibilities of action according to what Verbeek defines in terms of a certain "directedness" (Verbeek 2011, p. 15). In this specific case, the firearm makes it possible to shoot bullets, to attack or possibly defend oneself against potential attackers. In this sense, while one can certainly use a firearm for the purpose of, say, protecting the defenceless from a terrorist attack, it cannot be denied that the firearm is designed from the outset to shoot projectiles and that this use is anything but morally neutral. The firearm is thus a good example of how artefacts incorporate values[6] and how they mediate our experience of the world and our perception of ourselves.

Similarly, one can observe that the overpasses designed by Moses themselves represent further proof that artefacts are not morally neutral. Indeed, although it is not possible to identify values in some material part of the artefact for the simple reason that *values are not empirical*, and therefore the location criterion cannot be employed in the realm of artefacts nor in the exquisitely human realm, such artefacts do promote a form of racial or elitist discrimination in this case.

More specifically, it must be remembered that Moses deliberately designed the overpasses on the road leading to Long Island[7] to be so low that they could not be taken by buses, thus preventing access to certain beaches by anyone who could not afford to buy a car. In this way, Moses effectively excluded black and low-income people from certain areas of the city. By virtue of their configuration, such artefacts embody, according to Winner, relations of power or social order, and the fact that it is not possible to identify where the values reside does not detract from Winner's lucid examination.

In the same way, it can be observed that different technologies related to artificial intelligence incorporate different sets of values[8] (see, e.g., van de Poel 2020),

---

6 That a gun embodies values is also argued by Wallach and Allen, who in order to offer an example of operational morality call into question the childproof safety mechanisms with which certain weapons are equipped (Wallach and Allen 2009, p. 25).

7 Winner's example has been challenged by Woolgar and Cooper (1999).

8 The idea that technical artefacts incorporate values is nowadays the basis of the so-called Value Sensitive Design (VSD) approach. This approach aims to systematically integrate moral values into the design of technical artefacts. In other words, according to the Value Sensitive Design (VSD) approach we can somehow design technical artefacts in such a way that they can consciously become vehicles for moral values or at least avoid undesirable ethical fallout.

all of which implicate, in various ways, the different actors at work in the processes of design, production, and use of such technologies. Indeed, in addition to being capable of performing the function of *artificial moral advisor* (see, e.g., Giubilini and Savulescu 2018), with which we aim to consciously promote certain values within our society, such systems can also become a vehicle for discrimination and prejudice. In that respect, it suffices to mention the problem of biases, whereby intelligent systems have often led, and still lead, to results that reflect prejudices and bad practices that are widespread in society. Indeed, in some cases, algorithms reiterate and promote different forms of discrimination, thus demonstrating their ability to incorporate not only values but also disvalues. By virtue of this capacity, AI systems can in no way be considered morally neutral, as demonstrated, for instance, by the extensive debate that has begun in recent years on the principles and value frameworks to which algorithms themselves should be aligned (see Gabriel 2020). Although five principles (beneficence, non-maleficence, autonomy, justice, explicitness) are generally recognised (Floridi and Cowls 2019) as the foundation of a common code of artificial intelligence, the axiological question still remains widely debated and deserving of in-depth investigation, which should take into account both the principles on the basis of which artificial intelligence is developed and its by no means neutral moral status. We shall devote the next section to this latter issue in an attempt to make a contribution to clarifying the question of the moral agency[9] of artificial intelligent systems.

The moral significance of artificial intelligence: the problem of intentionality

From the broad and multi-faceted debate surrounding the moral status of artificial agents, we shall present here a few paradigmatic positions in order to compare them in the next section with Verbeek's postphenomenological theory.

The common thread running through these positions in the debate is the attribution (or non-attribution) of some form of *intentionality* to technical artefacts that would make it possible to consider them part of the moral world (or exclude them from it). The problem of intentionality is in fact a cornerstone in the construction of the notion of agency, as demonstrated by the fact that technological objects have so far been excluded from the field of ethics precisely because they lack intentionality and autonomy or because they have been reduced to mere mo-

9 We will not deal here with the problem of AI as a patient, to which a forthcoming paper of ours is devoted.

rally neutral instruments at the service of optimising outcomes. A solution to the problem of technological intentionality can thus help to redefine the moral significance of technical artefacts, and particularly those equipped with artificial intelligence.

In this regard, a particularly noteworthy reflection is the one put forward by Johnson in *Computer Systems: Moral Entities but Not Moral Agents* (2006). In this paper, the author aims to recognise a moral status in computer systems without attributing to them any form of *moral agency.* To accomplish this, Johnson first of all emphasises how computers do not possess certain characteristic traits of human agency – namely, mental states and intentions to act – and for this reason cannot be considered moral agents. This lack of consciousness and intention to act does not, however, lead the author to embrace the opposite thesis that these technological devices are morally neutral (see Johnson 2006, p. 195). Between these two opposites, Johnson ushers in a third possibility by introducing the category of *moral entities* to the debate. With this category, the author intends to affirm that computers are part of the moral world because they possess intentionality and efficacy by virtue of the way they are designed by humans and the functions they perform, even though they are not endowed with moral agency. Moral agency, in fact, requires a certain degree of freedom from which intentions to act can arise, which computers lack.

Now, the category of moral entities has the undoubted merit of highlighting what Ihde, as seen above, would define in terms of the primitive unit of man-technology, from which, for Johnson, the intentionality of technical artefacts would arise. Indeed, Johnson observes that artefacts only have moral significance in relation to the human being, i.e. as part of socio-technical systems, and it is as part of such systems that they present their own intentionality. More precisely, the intentionality of artificial entities emerges within a relationship involving the intentionality of programmers and users, in which the former is so to speak incorporated by the system, while the latter provides the input for the system's intentionality to be activated. The system's intentionality is thus identifiable with the capacity to provide output (the resulting behaviour) from given inputs, although without the programmers ever having specified the correlation between each peculiar input and each peculiar output:

> The system designer designed the system to receive input of a certain kind and transform that input into output of a particular kind, though the programmer did not have to specify every particular output for every possible input. In this way computer systems have intentionality [...] The intentionality of computer systems and other artifacts is connected to two other forms of intentionality, the intentionality of the designer and the intentionality of the user [...] The intentionality of computer systems is inert or latent without the intentionality of

> users. Users provide input to the computer system and in so doing they use their intentionality to activate the intentionality of the system (Johnson 2006, p. 201)

A second position that attributes intentionality to artificial agents is the one presented by John P. Sullins in *When Is a Robot a Moral Agent?* (2006). In this short paper, Sullins, who bases his reflections on the method of levels of abstraction proposed by Floridi and Sanders (2004), identifies three necessary requirements for a subject to be granted *full moral agency:* autonomy, intentionality, and responsibility.

As far as the first requirement is concerned, the author employs the engineering notion of autonomy without excessive precaution[10]. By virtue of this notion, a machine that presents a certain degree of independence with respect to other agents ("the machine is not under the direct control of any other agent or user" – Sullins 2006, p. 28) is therefore autonomous.

The second requirement, the subject of our examination, calls into question a weak notion of intentionality. In fact, although Sullins believes that it is possible to attribute intentionality to robots on the basis of the moral relevance of their actions, which, at a certain level of abstraction, may appear as calculated and deliberate, i. e. arising from *autonomous intentions*, this does not mean, however, that such artefacts should be recognised as having any form of intentionality in the strict sense, since – Sullins observes – this cannot be attributed to humans either. In this sense, "as long as the behaviour is complex enough that one is forced to rely on standard folk psychological notions of predisposition or 'intention' to do good or harm, then this is enough to answer in the affirmative to this question [concerning intentionality]" (Sullins 2006, p. 28). In other words, it is exclusively the behaviour of robots and not some characteristic of theirs that compels us humans to ascribe intentions to their actions.

Now, although both positions recognise a certain intentionality to technical artefacts, thereby legitimately relocating such artefacts within the domain of morality, the above arguments show two divergent tendencies at work that circumscribe the debate around the moral status of artificial entities. Indeed, although Johnson and Sullins manage to avoid the difficulties involving notions such as free will and intention to act by using a strategy of redefining the concept of intentionality itself, a clear difference separating the perspectives under discussion must be emphasised: Sullins bases his reflections on the moral significance of the actions per-

10 On the problems related to the notion of autonomy with regard to artificial agents, we refer to Johnson and Noorman (2014) and Loh and Loh (2017).

formed by such agents[11], whereas Johnson intends to demonstrate how artificial entities have moral value *as components of human action*. In this sense, it can be observed that Sullins's reflections, in the wake of those presented by Floridi and Sanders (2004), aim to liberate the artificial agent from a certain anthropocentrism[12] that tends to establish an intrinsic connection between the moral agency of the artefact and that of the human or, in some cases, that demands their full overlap in order for some form of moral agency to be attributed to artefacts. On the other hand, it should be noted, however, that Sullins's position is open to various criticisms that point out that behind the objectivity claimed by the method of abstraction levels proposed by Floridi and Sanders and taken up by Sullins himself, there is already an ethical choice that would determine the qualifying criteria for agenthood (Gunkel 2017, p. 73). In other words, such a method would not possess the objectivity it has in the field of mathematics from which it is borrowed, since at the root of the choice of the criteria for qualifying agenthood there would already be a decision as to who can be part of the community of moral subjects and who is excluded[13].

On the other hand, Johnson's reflections have their limits, too: they are exposed to the criticism of anthropocentrism, so agency is only human and cannot

11 In this regard, Coeckelbergh correctly observes that "against this approach [the one shared by Floridi, Sanders and Sullins], it could be argued that these arguments confuse moral *relevance* of actions with moral agency. It is one thing to recognize that such animals [rescue dogs] and robots do morally relevant things; it is another to say that they therefore have moral agency, which – so this argument goes – only persons or humans can have" (2020, p. 156).

12 For the authors, in fact, moral philosophy remains "unduly constrained by its anthropocentric conception of agenthood" (Floridi and Sanders 2004, p. 350).

13 More precisely, Gunkel states that "the method of abstraction, although having the appearance of an objective science modeled on 'the discipline of mathematics'[…], has a political-ethical dimension that is neither recognized nor examined by Floridi and Sanders. Whoever gets to introduce and define the LoA occupies a very powerful and influential position, one that, in effect, gets to decide where to draw the line dividing 'us from them'. In this way, then, the method of abstraction does not really change or affect the standard operating presumptions of moral philosophy or the rules of its game. It also empowers someone or something to decide who or what is included in the community of moral subjects and who or what is to be excluded and left on the outside" (Gunkel 2017, p. 73). In addition to this, Gunkel notes that this method does not avoid misunderstandings and disagreements about the criteria for qualifying agenthood: one need only think of Sullins, who employs such a method and yet recognises different criteria for agenthood than those of Floridi and Sanders (Gunkel 2017, p. 73). Another weak point in Floridi and Sanders's theory is emphasised by Verbeek, who, although he appreciates their work, observes that there are artefacts, such as ultrasound imaging or Moses's bridges, which "do not meet Floridi and Sanders's criteria for agency – but they do actively contribute to moral actions and have impacts that can be assessed in moral terms" (Verbeek 2011, p. 50).

be attributed to other subjects[14]. Moreover, in Johnson's eyes, the very moral significance of entities depends on their being components of human action, on their being designed by humans and functional for human purposes, even though these entities present a certain degree of independence, which is, however, not sufficient for them to be assigned the status of "autonomous moral agents" (Johnson and Miller 2008, p. 127). They are and remain an "extension of human activity and human agency" (Johnson and Miller 2008, p. 127).

Faced with the limits displayed by these reflections, we intend here to present Verbeek's postphenomenological position, which, as we will try to show, has the merit of bringing the man-machine relationship back within the notion of composite agent, thus freeing his proposal from the limits to which anthropocentric positions are exposed, without, however, involving the method of levels of abstraction, about whose validity various objections have been raised[15]. Between these two extremes lies our proposal to define, in the wake of Verbeek's reflections, AI systems in terms of moral mediators endowed with a technological intentionality linked to composite intentionality.

Technological intentionality and composite intentionality: Rethinking the moral status of artificial intelligence from a postphenomenological perspective

Verbeek's philosophical approach (Verbeek 2011; 2008; 2008a; 2005) is aimed at highlighting the role played by technologies in shaping our habits, and consequently recognising their clear moral significance. To this end, Verbeek develops in an original way certain fruitful insights found in Latour (see, e.g., Latour 1993; 1994; 2002) and Ihde (see, e.g., 1979; 1990; 1993), and he focuses his investigation on the function of mediation that technologies perform in our everyday lives. In the philosopher's eyes, in fact, technological devices contribute to shaping our ex-

14 For Coeckelbergh, in contrast to Johnson's position, "Floridi and Sullins could reply then that this definition of moral agency is too anthropocentric, that such a metaphysical freedom [from which for Johnson arises the intention to act that computers lack] is not needed for moral agency, or that its conditions are already fulfilled in the relevant cases of artificial agents such as the rescue dogs" (2020, p. 156).

15 Though they appreciate Floridi and Sanders's proposal, important objections are also raised by Johnson and Miller 2008.

perience of the world with significant impacts on our actions and decisions. In this sense, technological objects are not "neutral 'intermediaries' between humans and world, but *mediators*" that "actively mediate this relation" (Verbeek 2005, p. 114). By virtue of this role played by technology in the human-world relationship, ethics has, in Verbeek's view, the duty to extend its scope of enquiry beyond the human sphere, welcoming also "nonhuman forms of agency" (Verbeek 2011, p. 17), on which Latour has shed new light with his Actor-Network Theory (see Redaelli 2022).

In order to further develop this new form of ethics that seeks to overcome the modern subject-object dichotomy, albeit without embracing the thesis of symmetry laid out by Latour[16], Verbeek makes use mainly of the postphenomenological approach inaugurated by Ihde. As just mentioned, this approach has the merit of focusing on the role played by technologies in our relationship to the world[17]. In fact, starting from concrete empirical case studies as the basis for philosophical reflection, postphenomenology identifies different types of human-technology-world relations, focusing on "*how, in the relations that arise around a technology, a specific 'world' is constituted, as well as a specific 'subject'*" (Rosenberger and Verbeek 2015, p. 31). For Ihde, indeed, and so for Verbeek, too, the human-world relationship does not involve pre-existing subjects and objects, but rather subjects and objects are created and are co-constituted, in the interplay between humans, technology, and the world (Verbeek 2011, p. 15). In this sense, Ihde appropriately defines his approach as relativistic (Ihde 1990), not because it puts forward some form of epistemological relativism (see Verbeek 2005, p. 122), but because it is an analysis of relations.

Crucial to this type of analysis is the notion of intentionality, which Verbeek defines in the phenomenological terms of "directedness of human beings toward reality", whereby human beings cannot, according to Husserlian dictates, "simply 'think' but always think *something*; they cannot simply 'see' but always see *something*" (Verbeek 2011, p. 55). However, if, on the one hand, this notion of intentionality is explicitly borrowed from phenomenology, on the other hand, one must observe the peculiar curvature it undergoes in postphenomenological reflection. This redefinition is closely linked to what has just been stated: intentionality does not

---

16 In contrast to the symmetry between actants established by Latour, for their part Rosenberger and Verbeek (2015, p. 19) state that "the postphenomenological approach [...] explicitly does not give up the distinction between human and nonhuman entities. Instead of symmetry it sees interaction and mutual constitution between subjects and objects".

17 In this sense, postphenomenology does not see "phenomenology as a method to *describe* the world, but as understanding the *relations* between human beings and their world" (Rosenberger and Verbeek 2015, p. 11)

indicate a direct relationship between subject and object, but rather a relationship increasingly mediated by technologies, where this mediation is the fountain, the *source* (Rosenberger and Verbeek 2015, p. 12) from which subjectivity and objectivity take shape in specific situations. In this sense, Verbeek can recognise that intentionality is distributed between humans and technologies, attributing to the latter not some intention to act but a directedness, a "directing role in the actions and experiences of human beings" (Verbeek 2011, p. 57).

Starting from this reconceptualisation of the intentional relation, Verbeek then emphasises how human intentionality shaped by technological devices can take various forms (see Verbeek 2008). Of these forms, we intend to focus on what Verbeek defines in terms of composite intentionality, convinced that such intentionality characterises the human-AI relationship, since, in this variant, "there is a central role for the 'intentionalities' or directedness of technological artifacts themselves, as they interact with the intentionalities of the human beings using these artifacts" (Verbeek 2011, p. 145). In order to show the fecundity of this notion in the field of artificial intelligence, it is first necessary to trace its features.

First of all, Verbeek observes that "when the 'directedness' of technological devices is added to human intentionality, a *composite intentionality* comes about: a form of intentionality that results from adding technological intentionality to human intentionality" (Verbeek 2011, p. 145; see also Verbeek 2008). Thus, we can state that composite intentionality arises where there is a 'synergy' between technological intentionality "toward 'its' world", understood as a certain directionality, and human intentionality "toward the result of this technological intentionality" (Verbeek 2011, p. 146). In other words, in the case of composite intentionality "humans are directed here at the ways in which a technology is directed at the world" (Verbeek 2008, p. 393).

It is important to note here that, in Verbeek's view, this type of technological intentionality opens up a reality that is only accessible to technologies and which, at the same time, enters the human realm through technological mediation. Following this line of thought, Verbeek, who cites Hooijmans's photographs as an example, then also defines this type of intentionality in terms of "augmented intentionality", since it gives rise to an *expanded form* of intentionality that possesses both a representational and a constructive function. In other words, such intentionality cannot only represent reality but can only constitute a reality that exists for human intentionality if it is combined with technological intentionality.

This notion of composite intentionality seems to be particularly suited to explaining the intentionality present in AI systems while at the same time helping to clarify their status as moral mediators. Indeed, compared to, so to speak, traditional technologies, artificial intelligence makes autonomous decisions, adapting its behaviour according to the conditions in which it operates. In this capacity composite intentionality is revealed as a result, not merely a summation, of the human-machine association. This directivity or intentionality is – as Johnson observes – connected to the man who designs the machine and uses it, but at the same time – as Verbeek correctly underlines – presents an emerging character with respect to human intentionality, both that of the programmers and that of the users[18]. This character is not, however, reducible merely to the independence of the artificial agent, which despite such independence remains, in Johnson's reflection, an extension of the human itself, but is linked to its ability to structure new forms of reality (otherwise not accessible to man) according to unexpected directions of action[19]. In this sense, if it is correct to focus attention on the human-AI systems' intermingling in order to reassess the machine's moral status, thus avoiding falling into positions that only seemingly assume an objective view from which

---

18 Johnson seems to deny the emergent character of technological intentionality when she states that "even when it learns, it learns as it was programmed to learn. [...] The fact that the designer and user do not know precisely what the artifact does makes no difference here. It simply means that the designer – in creating the program – and the user – in using the program – are engaging in risky behavior. They are facilitating and initiating actions that they may not fully understand, actions with consequences that they cannot foresee. The designer and users of such systems should be careful about the intentionality and efficacy they put into the world. [...] When humans act with artifacts, their actions are constituted by their own intentionality and efficacy as well as the intentionality and efficacy of the artifact which in turn has been constituted by the intentionality and efficacy of the artifact designer" (Johnson 2006, pp. 203–204). Against such a linking of technological intentionality to human intentionality, which takes no account of emergent forms of mediation, and more generally of multistability, one can observe with Verbeek that "for even though because of their lack of consciousness artifacts evidently cannot form intentions entirely on their own, their mediating roles cannot be entirely reduced to the intentions of their designers and users. If they could be, the intentionalities of artifacts would merely be a variant of what John Searle called "derived intentionality" (Searle, 1983), entirely reducible to human intentionalities. Quite often (...) technologies mediate human actions and experiences in ways that were never foreseen or desired by human beings" (Verbeek 2011, p. 57). However, this does not mean that technological intentionality is independent of human intentionality, but rather that this is "one component of the eventually resulting intentionality of the 'composite agent'" (Verbeek 2011, p. 58). In bringing to light this so-to-speak emergent character of intentionality, a central point of differentiation between the two positions can be observed. In this regard, Gunkel correctly observes, "although Johnson's 'triad of intentionality' is more complex than the standard instrumentalist position, it still proceeds from and protects a fundamental investment in human exceptionalism. Despite considerable promise to reframe the debate, Johnson's new paradigm does not look much different from the one it was designed to replace. Human beings are still and without question the only legitimate moral agents" (Gunkel 2017, p. 68).

19 Verbeek defines technologies as "*mediators* that actively help to shape realities" (Verbeek 2011, p. 46).

to address the problem of moral standing[20], it should nevertheless be emphasised that the intentionality of AI systems goes beyond the triangulation highlighted by Johnson, since technological intentionality is not reducible to mere functionality designed by humans and initiated by the inputs entered by users. Rather, the notion of technological intentionality is linked to a certain notion of 'freedom' or non-deterministic character, because not all actions of intelligent systems are predictable and comprehensible (see the black box problem), and so their mediating role is also linked thereto. In this sense, artificial intelligent systems are characterised by a dual dynamic: they incorporate human intentions in a material way and at the same time present "emergent forms of mediation" (Verbeek 2011, p. 127), giving rise to a composite intentionality of the highest degree.

In order to clarify the notion of composite intentionality, it will be useful in closing our discussion to offer an example of where such intentionality is at work. We refer here to a Generative Art AI programme called Midjourney. This AI tool creates images from a short descriptive text entered by the user. In this case, as in the case of Hooijmans's photographs, the technology produces a reality that would not be experienced by the human subject at all, if his intentionality, in this case represented *in the first instance* by the descriptive text, were not supplemented by the intentionality of the intelligent system. This joint action, this composite intentionality, is even more evident since Midjourney uses the Discord channel in such a way that the user can enter prompts and interact with a bot that produces updated images in real time so that the user has the opportunity to make changes while still producing a result of human-AI interaction that cannot be reduced to mere human intentionality.

In fact, it is precisely this joint human-machine action with artificial intelligence that allows us to reaffirm the relational character of the notion of composite intentionality, which can also be found, to some extent, in some European directives on artificial intelligence (see Pacileo 2020). While requiring continuous human oversight of intelligent systems, which of course include Human-Centred AI[21], these directives focus on the joint character of human-machine action and the integrative (and not substitutive) function performed by artificial intelligence

---

20 In fact, for Johnson, "while Floridi and Sanders suggest that any level of abstraction may be useful for certain purposes, my argument is, in effect, that certain levels of abstraction are not relevant to the debate about the moral agency of computers, in particular, those levels of abstraction that separate machine behavior from the social practices of which it is a part and the humans who design and use it" (Johnson 2006, p. 196).

21 It should be noted here that Verbeek's post-humanist position does not seem to be at odds with Human-Centred AI (HCAI), but rather moves in the same direction insofar as it recognises that technologies in general and AI amplify and augment rather than displace human abilities.

with respect to our capabilities. Thus, also in the legal sphere, we can encounter the form of composite intentionality that has its highest expression in artificial intelligence systems *as* moral mediators.

References

Coeckelbergh, M. (2014). The Moral Standing of Machines: Towards a Relational and Non-Cartesian Moral Hermeneutics. *Philosophy & Technology, 27,* 61–77. DOI 10.1007/s13347–013–0133–8

Coeckelbergh, M. (2020). *Introduction to Philosophy of Technology.* Oxford University Press.

Coeckelbergh, M. (2020a). *AI Ethics.* The MIT Press.

Daza, M. T., and Ilozumba, U. J. (2022). A survey of AI ethics in business literature: Maps and trends between 2000 and 2021. *Frontiers in Psychology, 13.* https://doi.org/10.3389/fpsyg.2022.1042661

Floridi, L. and Cowls, J., (2019). A unified framework of five principles for AI in society. *Harvard Data Science Review* 1(1).

Floridi, L., and Sanders, J. W. (2004). On the Morality of Artificial Agents. *Minds and Machines, 14,* 349–379. https://doi.org/10.1023/B:MIND.0000035461.63578.9d

Gabriel, I. (2020). Artificial Intelligence, Values, and Alignment. *Minds & Machines, 30,* 411–437. https://doi.org/10.1007/s11023-020-09539-2

Giubilini, A., and Savulescu, J. (2018). The Artificial Moral Advisor. The "Ideal Observer" Meets Artificial Intelligence. *Philosophy & Technology, 31,* 169–188. https://doi.org/10.1007/s13347-017-0285-z

Gunkel, D. J. (2017). *The Machine Question: Critical Perspectives on AI, Robots, and Ethics*, The MIT Press.

Johnson, D. G. (2006). Computer systems: Moral entities but not moral agents. *Ethics and Information Technology, 8,* 195–204. https://doi.org/10.1007/s10676-006-9111-5

Johnson, D. G., and Miller, K. W. (2008). Un-making artificial moral agents. *Ethics and Information Technology, 10,* 123–133. https://doi.org/10.1007/s10676-008-9174-6

Johnson, D. G., and Noorman, M. (2014). Artefactual agency and artefactual moral agency. In P. Kroes and P.-P. Verbeek (Eds.), *The Moral Status of Technical Artefacts* (143–158). Springer Science+Business Media.

Ihde, D. (1979). *Technics and Praxis.* Reidel.

Ihde, D. (1990). *Technology and the Lifeworld: From Garden to Earth.* Indiana University Press.

Ihde, D. (1993). *Postphenomenology.* Northwestern University Press.

Latour, B. (1993). *We Have Never Been Modern*, eng. trans. by Catherine Porter, Harvard University Press.

Latour, B. (1994). On Technical Mediation: Philosophy, Sociology, Genealogy. *Common Knowledge, 3*(2), 29–64.

Latour, B. (2002). Morality and Technology: The End of the Means. *Theory, Culture and Society,* 19, 247–260.

Loh, W., and Loh, J. (2017). Autonomy and Responsibility in Hybrid Systems: The Example of Autonomous Cars. In P. Lin, K. Abney, and R. Jenkins (Eds.), *Robot Ethics 2.0: From Autonomous Cars to Artificial Intelligence* (pp. 35–50). Oxford University Press. https://doi.org/10.1093/oso/9780190652951.003.0003

Millar, J. (2017). Ethics Settings for Autonomous Vehicles. In P. Lin, K. Abney, and R. Jenkins (Eds.), *Robot Ethics 2.0: From Autonomous Cars to Artificial Intelligence* (pp. 20–34). Oxford University Press. https://doi.org/10.1093/oso/9780190652951.003.0002

Pacileo, F. (2020). L'uomo al centro. IA tra etica e diritto nella responsabilità d'impresa. in M. Bertolaso, G. Lo Storto (Eds.), *Etica Digitale. Verità, responsabilità e fiducia nell'era delle macchine intelligenti* (83–99). Luiss University Press.

Peterson, M. and Spahn., A. (2011). Can Technological Artefacts Be Moral Agents? *Sci Eng Ethics*, *17*, 411–424. https://doi.org/10.1007/s11948-010-9241-3

Pitt, J. C. (2014). "Guns Don't Kill, People Kill"; Values in and/or Around Technologies. In P. Kroes and P.P. Verbeek (Eds.), *The Moral Status of Technical Artefacts* (89–101). Springer Science +Business Media. DOI:10.1007/978-94-007-7914-3_6

Redaelli, R. (2022). Composite Intentionality and Responsibility for an Ethics of Artificial Intelligence. *Scenari*, *17*, 159–176.

Rosenberger, R. and Verbeek, P.P. (2015). A Field Guide to Postphenomenology. In R. Rosenberger, and P.P. Verbeek (Eds.), *Postphenomenological Investigations: Essays on Human-Technology Relations* (9–41). Lexington Books.

Sparrow, R. (2016). Robots in aged care: A dystopian future? *AI and Society*, *31*(4), 445–454. https://doi.org/10.1007/s00146-015-0625-4

Sparrow, R., and Sparrow, L. (2006). In the hands of machines? The future of aged care. *Minds & Machines*, *16*, 141–161. https://doi.org/10.1007/s11023-006-9030-6

Sullins, J. P. (2006). When is a Robot a Moral Agent? *International Review of Information Ethics*, *6*, 23–30.

van de Poel, I. (2020). Embedding Values in Artificial Intelligence (AI) Systems. *Minds & Machines*, *30*, 385–409. https://doi.org/10.1007/s11023-020-09537-4

Verbeek, P.P. (2005). *What Things Do: Philosophical Reflections on Technology, Agency, and Design.* Penn State University Press.

Verbeek, P. P. (2008). Cyborg intentionality: Rethinking the phenomenology of human-technology relations. *Phenomenology and the Cognitive Sciences*, *7*(3), 387–395. https://doi.org/10.1007/s11097-008-9099-x

Verbeek, P. P. (2008a). Obstetric Ultrasound and the Technological Mediation of Morality: A Postphenomenological Analysis. *Human Studies*, *31*, 11–26. https://doi.org/10.1007/s10746-007-9079-0

Verbeek, P. P. (2011). *Moralizing Technology: Understanding and Designing the Morality of Things.* University of Chicago Press.

Wallach, W., and Allen, C. (2009). *Moral Machines. Teaching Robots Right from Wrong.* Oxford University Press.

Winner, L. (1986). Do Artifacts Have Politics? In *The Whale and the Reactor*, University of Chicago Press.

Woolgar, S. and Cooper, G. (1999). Do Artefacts Have Ambivalence? *Social Studies of Science*, *29*(3), 433–449.

Bei Fragen zur Produktsicherheit wenden Sie sich bitte an:
If you have any questions regarding product safety,
please contact:

Walter de Gruyter GmbH
Genthiner Straße 13
10785 Berlin
productsafety@degruyterbrill.com